A PREFACE TO MARK

A PREFACE TO MARK

Notes on the Gospel in Its Literary and Cultural Settings

CHRISTOPHER BRYAN

New York Oxford
OXFORD UNIVERSITY PRESS
1993

Oxford University Press

Oxford New York Toronto
Delhi Bombay Calcutta Madras Karachi
Kuala Lumpur Singapore Hong Kong Tokyo
Nairobi Dar es Salaam Cape Town
Melbourne Auckland Madrid

and associated companies in
Berlin Ibadan

Published by Oxford University Press, Inc.
200 Madison Avenue, New York, New York 10016

Library of Congress Cataloging-in-Publication Data
Bryan, Christopher, 1935–
A preface to Mark : notes on the Gospel in its literary
and cultural settings /
Christopher Bryan.
p. cm.
Includes bibliographical references and index.
ISBN 0-19-508044-0
1. Bible. N.T. Mark—Introductions.
I. Title.
BS2585.2.B79 1993 226.3'06—dc20 92-40917

2 4 6 8 9 7 5 3 1

Printed in the United States of America
on acid-free paper

In Memoriam

SPLODGER
1974–1991

POOH
1975–1992

Faithful, Loving, and Beloved Friends

צִדְקָתְךָ כְּהַרְרֵי-אֵל מִשְׁפָּטֶיךָ תְּהוֹם רַבָּה
אָדָם וּבְהֵמָה תוֹשִׁיעַ יהוה:
תהלים לז

Preface

Prefatory notes and acknowledgments have their perils. Not least among these is the likelihood that those acknowledged may be more embarrassed by association with one's meanderings than they would have been hurt by silent ingratitude. Nonetheless, some things have to be said.

I should like to express my thanks to Professor David Catchpole of the University of Exeter for his friendship, help, and many kindnesses to me during my sabbatical year in Exeter 1989–90, and since. I wish also to thank the Reverend Richard Burridge, Chaplain to the University of Exeter, for his friendship and help, and in particular for drawing my attention to his masterly doctoral thesis on the subject of the gospels and Greco-Roman biography, subsequently published by Cambridge University Press (1992). I must thank the librarians of the University of the South, Edward Camp, Sue Armentrout, Donald Haymes, and Charles Van Hecht; and my research assistants over several years, the Reverend Lorraine Ljunggren, the Reverend Scott Lee, and Susan Bear, Joyce Latimer, Amy Carol Bentley, Sarah Gaede, and Ramiro Eduardo Lopez. All have assisted me in more ways than I can name. I must thank the Reverend Ellen B. Aitken, now studying at Harvard, and formerly my research assistant, who several years ago first drew my attention to the researches of Milman Parry and his colleagues and successors. I am grateful to the American Association of Theological Schools for a generous grant during my sabbatical year. I am grateful to Cynthia Read, Susan Hannan, and their colleagues at Oxford University Press for constant encouragement and invaluable criticism as my manuscript ground slowly toward its final state. I must thank countless colleagues, friends, students, and teachers who, over the years, have assisted and enriched my thoughts about the gospel in more ways than I can possibly remember.

Finally I thank Wendy Bryan for the good nature and patience with which she has lived with this study for nearly three years. Last summer we went to see Alan Ayckbourn's powerful and disturbing comedy, *Woman in Mind*. The play centers on the experience of a woman unfortunate enough to be married to a parson whose life is dominated by production of what he refers to repeatedly as "the book." No one, I suspect, will be surprised to learn that I watched with more than a twinge of guilt and embarrassment.

I originally intended to dedicate my work to Wendy, and if there is any virtue in such a dedication, no one could deserve it more. I changed my mind for reasons that she, at least, will understand. Throughout most of my writing I was accompanied by our mongrel bitches Splodger and Pooh. Splodger would sit on the green rug in the study and supervise. She was always ready to urge me to renewed activity if I seemed to be taking too long over a meal or a coffee break. Pooh, meanwhile, was content to sleep on the sofa *except* at meal and coffee-break times, since she regarded these as the most important part of the day's activity. Alas, while the closing chapters were being revised, Splodger died. Something less than a year later her inseparable friend Pooh, who had grieved for her perhaps more than any of us, died too. They were dogs of infinite character, each game to the last, and they are sorely missed. I dedicate this book to their memory, presuming (as it is written) that,

> Thy righteousness standeth like the strong mountains:
> Thy justice is like the great deep:
> Humankind and beast shalt thou save, O Lord.
>
> (Psalm 36.6)

Sewanee, Tenn.
December 1992

C. B.

Contents

Part II Was Mark Written to Be Read Aloud?

A PREFACE TO MARK

Prologue: Looking at Mark

Scope of the Inquiry

The Gospel according to Mark[1] is a Greek text composed in the second half of the first century of the Christian era. It was written at a time when authors who wrote for publication generally expected their work to be read aloud in groups, rather than privately or silently by individuals. If these assertions are accepted, at least provisionally, then we are offered two areas of inquiry in our attempt to see Mark in its literary and cultural setting. First, we can reflect on the fact that Mark is a text, and ask what kind of text it would have been seen to be, either by its author or by others who encountered it near to the time of its writing. Should Mark be seen as an example of any particular literary type, and if so, which? Here we shall need to compare Mark with other texts of the period. This inquiry will occupy us in Chapters 2 to 9.

Second, we may ask whether Mark, like so much else of its period, was written to be read aloud. If so, how could we tell? What would be the signs of such an intention? Here we shall need to consider differences between approaches to composition and narrative oriented toward listeners, and those oriented toward readers, and to ask whether Mark is illuminated in the light of this. Are there ways in which Mark's

1. It will be convenient throughout this study to speak of the author of the second gospel as "Mark." As it happens, what follows does not appear to militate against traditional views of authorship and provenance; neither, however, does it confirm them.

narrative would have worked particularly well as rhetoric? These inquiries will occupy us in chapters 10 to 16.

Purpose of the Inquiry

To what end do we ask our questions? Broadly, to see how the evangelist's work is likely to have been regarded and understood in its original setting. What, to put it another way, could those who first encountered the gospel have been expected to make of it? I do not reckon that the answer to that question is entirely to be divorced from the answer to another, namely, What did the evangelist intend? although certainly I concede that neither the value of a text nor its significance are exhausted by or equivalent to its author's intention. The value of any work of art stems finally not from what its author meant, but from the response it evokes in us. Those works that affect us most and best we call inspired.

BIBLIOGRAPHY

The best general commentaries available in English on Mark are probably still Vincent Taylor, *The Gospel According to St. Mark* (London and New York: Macmillan, 1952) and C. E. B. Cranfield, *The Gospel According to Saint Mark,* 3d ed., with supplementary notes (Cambridge: Cambridge University Press, 1966). Among recent commentaries, Morna D. Hooker, *The Gospel According to St. Mark* (London: A. and C. Black, 1991) is straightforward and sensible.

There has been an abundance (perhaps even a plethora) of books about Mark published in the last decade. Many approach the text with questions and concerns quite different from those I have proposed. Some, however, relate directly to my areas of interest. Among these, Vernon K. Robbins's *Jesus the Teacher: A Socio-Rhetorical Interpretation of Mark* (Philadelphia: Fortress, 1984) and Mary Ann Tolbert's *Sowing the Gospel: Mark's World in Literary-Historical Perspective* (Minneapolis, Minn.: Fortress, 1989) are major works. Shorter and less detailed, yet offering a wealth of insights into Mark's construction and literary methods, are Joanna Dewey, *Markan Public Debate: Literary Technique, Concentric Structure, and Theology in Mark 2:1–3:6,* Dissertation series 48 (Chico, Calif.: Scholars, 1980); Augustine Stock, O.S.B., *Call to Discipleship: A Literary Study of Mark's Gospel* (Wilmington, Del.: Michael

Glazier, 1982); and Paul Achtemeier, *Mark* (Philadelphia: Fortress, 1986). Among earlier studies, Robert H. Lightfoot, *The Gospel Message of Saint Mark,* 2d ed., corrected (Oxford: Clarendon, 1952) is another valuable resource. On the questions of Markan provenance and authorship, beside the usual commentaries, one can hardly do better than Raymond E. Brown and John P. Meier, *Antioch and Rome: New Testament Cradles of Catholic Christianity* (New York and Ramsay, NJ: Paulist, 1983), 191–202.

I

WHAT KIND OF TEXT IS MARK?

1

The Question of Genre

The Significance of Genre

"The first qualification for judging any piece of workmanship from a corkscrew to a cathedral is," observed C. S. Lewis, "to know *what* it is—what it was intended to do and how it is meant to be used."

> After that has been discovered the temperance reformer may decide that the corkscrew was made for a bad purpose, and the communist may think the same about the cathedral. But such questions come later. The first thing is to understand the object before you: as long as you think the corkscrew was meant for opening tins or the cathedral for entertaining tourists you can say nothing to the purpose about them (Preface to *Paradise Lost*, 1).

What then is, or was, the gospel of Mark? What kind of text would it have been seen to be, either by its author or by its first readers and auditors? We are speaking of what most literary critics refer to as the problem of "genre," though personally I should have preferred "type" or "kind," or even Lewis's Latin "genus" (which could be conveniently contrasted with "species"). Be that as it may, the history of discussions of literary genres is a long one, and the nature and significance of literary genre remain matters of debate. What follows merely sketches one view of these questions.

In a recent interview, the novelist Saul Bellow observed that if we are truly to understand literature, we need what he calls a "trained sensibility." How do we develop such a sensibility? It is, said Bellow, impossible, "unless you take certain masterpieces into yourself; as if you were swallowing a communion wafer. If masterpieces don't have a decisive part in your existence, all you have is a show of culture. It has no reality" (Botsford, "Saul Bellow," 29). This is the essence of classicism. As Henri Marrou has powerfully shown us, the παιδεία (transliterated *paideia*) (education) of Hellenistic antiquity was at its best a system whereby persons were valued not for their wealth or power, but for the degree to which they had developed the fullest potential of their minds and their bodies—in later Hellenism, particularly their minds. This development was not directed toward what we would call "general education," but rather to what were regarded as the highest achievements and the best guides available for anyone who would be truly human, namely, the literature of Hellenism itself. Homer and the other classical writers formed virtually a "canon" of what the educated person should know. "So always read authors who are tried and tested [*probatos itaque semper lege*]," said Seneca (*To Lucilius* 2.4). Modern observers have sometimes regarded this concern with the classics as backward-looking and uncreative, but that is entirely to misunderstand it. Hellenism was not backward-looking. It simply recognized a certain group of masterpieces as embodying its values. When those masterpieces had been produced was of no importance. What was important was they were *there*, to be admired and rediscovered for all time.

It is inevitable, however, that in reflecting upon great works of art one should sooner or later ask, "How does the thing work?" So it was that in the course of discussing and observing these masterpieces there arose naturally the notion of certain "rules" governing what was proper to the various kinds of literature (for example, Aristotle, *Poetics* 1447a 8–13), rather as dogmatic theology and articles of religion arose in discussion and observation of faith and worship. Hence classical literary criticism was typically concerned with propriety of language, propriety of subject, and other ways by which a particular piece of literature might be seen to have conformed, or not to have conformed, to its type. Modern observers have sometimes accused classical criticism of narrowness because of this concern, but that is, again, entirely to misunderstand it. Just as no real theologian ever confused dogmatic theology or articles of religion with faith and worship, neither did any real literary critic ever confuse "the rules" with that which finally makes literature valuable to us. The Latin poet and critic Horace (65–8 B.C.E.)

was deeply concerned with propriety (*decorum*), but he was also clear that "where the beauties in a poem are many, I shall not be offended by a few blots" (*Art of Poetry* 351–52; cf. 1–9). And the anonymous critical genius known to history as "Longinus" (probably early second century C.E.) reminded fellow students and scholars that "the sublime" is beyond critical regulation:

> Other qualities show that their possessors are human; sublimity brings them near to the magnanimity of God. What is correct avoids censure; but what is great excites our awe. We need scarcely add that each of these great ones time after time redeems all errors by a touch of sublimity and true excellence. What is ultimately decisive is this, that if we were to pick out every fault in Homer, Demosthenes, Plato, and all the other greatest authors and combine them, we should find them to be a minute fraction, not the tiniest part, of the real excellence that is found in every part of these demi-gods. That is why the judgment of every age, which no envy can show to be in error, has bestowed upon them the prizes of victory. (*On the Sublime*, 36)

What then is the significance of classical interest in the "rules" of genre? Simply that a feeling for *type*, a sense of what *kind* of book or work of art this is, effects both how an artist proceeds, and how a resulting work is received. If we pick up a novel having been led to expect that it will be a detective story, and it turns out to be a serious study of personal relationships, we may well experience some difficulty adjusting to it. We cannot, as we say, "get into" it. We may even feel cheated. When Milton hesitated between epic "following the rules of Aristotle" (that is, epic following the precept of unity of action, which he eventually chose for *Paradise Lost*) and epic "following nature" (that is, epic with multiplicity of action, such as we find in the romantic epics of Boiardo and Spenser), he was hesitating between different kinds of expectation in his audience, different kinds of pleasure, and different achievements. These differences are essentially what the classical critics had discerned, and attempted to codify. Genre is, as Alastair Fowler has expressed it, "a communication system, for the use of writers in writing, and readers and critics in reading and interpreting" (*Kinds of Literature*, 256). Our notions of genre set boundaries between authors and audience, and begin to clear the way for an audience to perceive an author's purpose.

There is thus real value in our discerning what genres writers are using. Naturally, artists stretch and remold genres all the time—sometimes too quickly for public taste; frequently, writers are misunderstood in their own time and only later are highly regarded. On the other hand, the notion of a writer proceeding *without* genre, or creating a totally new genre (*sui generis*), is (even if theoretically possible) akin to the notion of a writer choosing to write in an unknown language. It is the notion of a writer choosing to be incomprehensible. In fact, the wildest artistic experimentation invariably has some connection, either by adaptation or reaction, to what has preceded it.

On Recognizing Genre

How do we recognize the genre of a particular work of art? By identifying a whole series of elements. When the genre is familiar to us, we do this effortlessly, recognizing and decoding almost without thought. Consider, for example, a rather simple twentieth-century genre, recently become somewhat passé: the "western." The chances are that even the title of such an entertainment (*Wagon Train*, *Gunfight at the O.K. Corral*) gave us some clue as to what it would be about: certainly the artwork that advertised it did. As the performance began, the credits often unfolded against settings of big skies and open country. Possibly there were shots of a lone horseman wearing a Stetson, or covered wagons crossing the prairie. All confirmed our expectations. As the drama unfolded, we undoubtedly recognized various themes and motifs: the good man alone (or with a few chosen companions) who fought for justice, the bad men who terrorized the town, the saloon girl with a heart of gold, the town drunk who made good in the end, the final shootout, and many more. Of course not every western contained every motif. *High Noon* had no town drunk and no saloon girl with a heart of gold. What is more, *High Noon* contained a (for the period) rather careful study of the relationship between the hero and the heroine, a motif not traditionally associated with westerns at all, but with romantic drama. Despite these variations, the western motifs that *were* present in *High Noon* were dominant, and we were never in any doubt as to what we were watching. Indeed, aficianados of the genre even refer to *High Noon* as a "classic" western. It is, again, fairly unusual for a "western" of the "U.S. Cavalry" type (a well-established subgenre) to end with the death of the hero and his regiment; but that is exactly what happened at the end of *They Died With Their Boots*

On—in this case, whatever the propensities of the genre, its use was controlled by the tradition. Even Hollywood dared not pretend that General George Custer and the Seventh Cavalry survived the Little Big Horn. So Hollywood, consciously or not, borrowed a motif from a folklore more ancient than that of the American West: the Last Stand of the Hero and his Companions.

> Hige sceal þe heardra, heorte þe cenre,
> mod sceal þe mare, þe ure mægen lytlað.
> (*Battle of Maldon* 312–13).
>
> [Resolve shall be the firmer, heart the braver,
> Spirit the greater, though our might lessens.]

The result is possibly the most memorable part of the movie.

This simple example illustrates much about the nature of genre. First, genre involves a *cluster* of elements. So striking are these elements that we can entirely understand why one might be tempted to regard them as "rules." Yet they are not precisely "rules," for they need not all be present in any one example. The genre of a particular work is established by the presence of *enough* generic motifs in sufficient force to dominate. Second, a work of one genre may contain motifs from another. This means that in establishing genre we need to identify the *dominant cluster* of motifs: just one or two will not do. Thus it may be the case, as some critics have suggested, that the written gospel has elements in common with drama or rhetoric. Such elements alone, however, do not make the written gospel an example of either, any more than the love interest in *High Noon* makes it into a romantic drama, or the ribald Porter's scene in *Macbeth* makes that into a comedy. There are other elements in the written gospel that may reasonably be described as midrash—using the word "midrash" in that broad sense whereby, as Jacob Neusner puts it, it "stands for pretty much anything any Jew in antiquity did in reading and interpreting Scripture" (Neusner, *What Is Midrash?* xii). But that, again, does not mean that the written gospel is in any way formally or generically a midrash, as will be clear from the the briefest comparison with classic midrashim such as *Mekhilta*, *Midrash Rabbah*, or *Pesikta deRav Kahana*.

Like everything else, the admixture of motifs from one genre into another can be done well or ill, successfully or unsuccessfully. Shakespeare's *The Winter's Tale* ends as romantic comedy, yet many critics have observed that the dramatist's use of tragic motifs in the first three acts is so powerful and so sustained as nearly to undermine the whole thing. By contrast, it is arguably just the *non*western motifs in *High Noon* and *They Died With Their Boots On* that offer the greatest interest. Some genres, indeed, seem naturally to overlap. A biography, for example, may tend toward history (a coincidence noted by the ancients as well as by modern students of the genre), encomium, hagiography, satire, or even the novel.

BIBLIOGRAPHY

On the Hellenistic *paideia* generally, Henri I. Marrou, *A History of Education in Antiquity* (London: Sheed and Ward, 1956) is foundational. See further M. L. Clarke, *Higher Education in the Ancient World* (London: Routledge and Kegan Paul, 1971), and Stanley F. Bonner, *Education in Ancient Rome: From the Elder Cato to the Younger Pliny* (Berkeley and Los Angeles: University of California Press, 1977). On classical literary criticism, see George A. Kennedy, ed., *Cambridge History of Literary Criticism*, vol. 1, *Classical Criticism* (Cambridge: Cambridge University Press, 1989). For a negative view of classical approaches to literature, see J. D. Denniston, *Greek Literary Criticism* (London: Dent; New York: Dutton, 1924), xxxiii–xxxvi.

The most important modern discussion of genre is probably Alastair Fowler, *Kinds of Literature: An Introduction to the Theory of Genres and Modes* (Oxford: Clarendon, 1982). See also Fowler's earlier essay, "The Life and Death of Literary Forms" in Ralph Cohen, ed., *New Directions in Literary History* (London: Routledge and Kegan Paul; Baltimore, Md.: Johns Hopkins University Press, 1974). Compare also René Wellek and Austin Warren, *Theory of Literature*, 3d ed. (Harmondsworth, Middlesex: Penguin, 1963), 18–19, 226–37, and E. D. Hirsch, *Validity in Interpretation* (New Haven: Yale University Press, 1967), 68–126.

On the gospel as drama, see B. H. M. G. M. Standaert, *L'Evangile selon Marc: Composition et Genre Littéraire* (Zevenkerken: Brugge, 1978); also Gilbert B. Bilezikian, *The Liberated Gospel: A Comparison of the Gospel of Mark and Greek Tragedy* (Grand Rapids, Mich.: Baker, 1977). There are useful general observations in Augustine Stock, O.S.B., *Call to Discipleship: A Literary Study of Mark's Gospel* (Wilmington, Del.: Glazier, 1982), 16–30.

For a view of a written gospel as midrash, see Matthew Goulder, *Midrash and Lection in Matthew* (London: SPCK, 1974); for criticism, see Philip S. Alexander, "Midrash and the Gospels," in C. M. Tuckett, ed., *Synoptic Studies: The Ampleforth Conferences of 1982 and 1983*, JSNT Supplement Series 7 (Sheffield: JSOT Press, 1984), 1–18; also Brian McNeil, "Midrash in Luke?"

Heythrop Journal 19 (1978): 399–404. For a general introduction to midrash, see Jacob Neusner, *What is Midrash?* (Philadelphia: Fortress, 1987) and *Invitation to Midrash: The Workings of Rabbinic Biblical Interpretation* (San Francisco, Calif.: Harper and Row, 1989). For further discussions of midrash in senses so broad that the word seems virtually indistinguishable from "exegesis" or "interpretation," see the nevertheless rich and fascinating *Midrash and Literature*, ed. Geoffrey H. Hartman and Sanford Budick (New Haven: Yale University Press, 1986).

2

Mark's Milieu

The recognition of a genre is, then, easy when we are familiar with it, and that is why genres are useful. But genres change, and particular genres belong to particular milieus, so problems arise when the text we are considering is two thousand years old. How can we identify the texts with which contemporaries might have compared or contrasted Mark? Where should we look for evidence of his literary milieu? The answers require critical and historical judgment, and involve the uncertainties and ambiguities of any subjective procedure. Nevertheless, there are some criteria that may fortify us against mere caprice, notably criteria of place, social level, and time.

As regards place, it is evident that throughout the first century of the Christian era the dominant culture, which we call Hellenism, was international, at least for those countries surrounding the Mediterranean basin. The triumph of Hellenism was above all the triumph of a particular pattern of *paideia*. The same books were read by educated persons from Spain to Libya. The same canons of taste and intellect were admired everywhere. Since the gospel of Mark has rather obvious Jewish connections, we should note that Judaism, too, participated in this culture, and had done so for centuries. Not merely the Hellenist Philo of Alexandria, but even Palestinian rabbis were influenced by Greek language and Greek literary and intellectual method; the Jewish Scriptures in their turn could, in Greek dress, rouse the interest and admiration of gentile literary critics (Longinus, *On the Sublime* 9.9). Indeed it was, as Jacob Neusner has pointed out, the very prevalence of this shared culture that made it so important for Jews to define and claim their distinctive identity (*Method and Meaning*, 22). So then, provided that our literary evidence is not clearly tied to a particular geographic locality, we may in general risk drawing it from wherever within the Greco-Roman world we find it.

As regards social level, certainly we must always be aware of distinctions between the learned and the simple, the aristocratic and the popular. Only comparatively recently has scholarship begun to take seriously that beside the more sophisticated literary products of the ancient world there also existed a considerable popular production aimed at a large and growing band of literacy in the middle layers of society, including both men and women: well-to-do traders, artisans, administrators, and clerks. Needless to say, types of literature that appealed to the intellectual and social elite have been better preserved than these more popular texts; it will always be easier to find a year-old copy of the *New York Times* or the London *Independent* than a year-old copy of the *National Enquirer* or the *Sun*. Yet some popular texts from antiquity have been preserved, such as Chariton's *Callirhoe* (a romantic novel, probably from the mid-first century C.E.), Xenophon's *Ephesian Tale* (early or mid-second century, also a romantic novel), and the anonymous *Life of Secundus the Silent Philosopher* (late second century) (see Appendix). Since those middle and working classes at whom such texts were aimed were, as regards their social and cultural level, precisely those who seem also to have formed the bulk of the early Christian communities, we shall do well not to ignore them in our reflections on the literary provenance of the written gospel, bearing in mind that Origen and other Christian apologists found themselves obliged to defend the simplicity and popular style of Christian writings (Origen, *Against Celsus* 1.62; Isidore of Pelusium, vol. 4, epistle 67; cited in Voelz, 895–96).

On the other hand, that distinction granted, we should not make it into an absolute, or speak as if there were no gradations between the elite and the vulgar. A taste for Jane Austin and Tennessee Williams does not necessarily exclude a taste for Dick Francis and Dorothy Sayers. There are those among the learned who have been known not only to enjoy popular thrillers and fantasies, but even to write them. No doubt the intellectual elite of the Greco-Roman world also included some who could suspend their sophisticated literary canons for the sake of a good tale.

As it happens, various aspects of life in the Hellenistic age suggest that it offered opportunities for a degree of culture in common between high and low such as were not to be found again until the advent of generally available electronic media in the twentieth century. On the most basic level, the extended family (Greek, οἰκία Latin, *familia*), which was the primary social unit in antiquity, provided a setting in which masters and mistresses, slaves and freedpersons, could not fail to affect and be affected by each other's thought and culture. Within Judaism, it cannot be doubted that the already widespread institution of the synagogue provided a further culturally leveling influence.

In the Hellenistic world at large, plays and poetry were normally performed publicly (without charge for admission) in theaters and lecture halls. Lectures and discussions took place in public colonnades. The administration of justice was a public act, if not a public entertainment, and Quintilian felt bound to warn would-be advocates that a case could be won or lost by applause from the crowd (*Institutio Oratoria* 8.3.3; compare 11.3.131). At events like the Olympic and Isthmian Games people from every walk of life would look to be entertained not only by jugglers and athletes, but also by talk from sophists, and by authors declaiming poetry and prose. Into such a scene, we may note, the New Testament's own reference to Paul teaching publicly in the hall of Tyrannus fits perfectly (Acts 19:9–10).

Of course general interest does not necessarily follow from general availablity. No doubt many shrugged after a few minutes of the poetry or the lecture, and went to watch the jugglers. But some would have found their attention caught, and stayed. The British public is not generally noted for its enjoyment of grand opera. In 1990, however, the British Broadcasting Corporation used a recording of Puccini's "Nessun dorma" by Luciano Pavarotti and the London Philharmonic Orchestra as theme music for its transmissions of soccer games from the immensely popular World Cup. Within weeks, sales of the recording, usually numbered in hundreds, had reached four hundred thousand. No one doubted a connection between the repeated transmissions and the sales.

For all these reasons, therefore, while we need to bear in mind social distinctions—noting, for example, that the thoughts and expressions of the aristocratic Philo or the learned scribes can hardly be transferred without remainder to the gatherings of a middle-class Christian household in Rome or Corinth—still we should not assume that the former cannot, if used with caution, help us to understand the latter, or that either may not help us to understand the milieu of the written gospel. No doubt literati differed from their academic inferiors in first-century Rome as in twentieth-century Europe or America, but they do seem to have talked about some of the same things, and even to have used some of the same words.

What, finally, of time? Here we must express a somewhat sterner caveat. Hellenistic culture was by our standards stable—indeed, by our standards, remarkably so—and it is a commonplace to say that Jewish tradition was conservative. In seeking the literary milieu of Mark we may therefore plausibly risk selection of documents from a much longer period than would be admissible if we were seeking to examine the milieus of eighteenth-, nineteenth-, or twentieth-century writers, with regard to whom we often define literary "periods" within decades (the "period" of the English Romantics, or even the "age" of Johnson!). Yet even in antiquity, time and events had their consequences. It is perfectly clear that the Fall of Jerusalem and the disastrous Bar Kochba War had profound and lasting effects upon Judaism; while Hellenism did not have to come to terms with anything quite so catastrophic as these, nevertheless, there were obviously changes and developments. For example, at some time between 62 and 65 C.E., Seneca is complaining that in his day the bright young things are no longer dressing modestly (*Epistle* 19.20). On the brighter side, about a generation later, Pliny, Tacitus, and Plutarch are all expressing a degree of optimism about society in general. Pliny speaks of revival in religion (*Letters* 10.96), Tacitus of a new freedom of expression (*Agricola* 3), and Plutarch claims that "wars have ceased, there are no wanderings of peoples, no civil strifes, no despotisms, nor other maladies and ills in Greece" (*The Oracles at Delphi no Longer Given in Verse* 408.b–c). But then, about a decade after his *Agricola*, we find Tacitus in the *Dialogue on Oratory* wrestling with the problem of a decline in Roman political oratory (*Dialogue on Oratory* 1). Some years after that, Lucian is chattering away to the effect that tragedies are not being performed any more and "absolutely no-one isn't writing history" (*How to Write History* 2). Now certainly we should be very foolish to draw any conclusions about the nature or depth of changes in Hellenistic society from such a mixed bag of reactions and comments as these. My point is simply that some of those who lived in the Hellenistic world during the first centuries of the Christian era thought that they perceived changes, and we should be unwise to ignore them.

Jacob Neusner has demonstrated that we cannot without great caution use later rabbinic texts as sources for information about the first century of the Christian era (for example, *Rabbinic Traditions*, passim; *Method and Meaning*, 185–213), and F. Gerald Downing (*A bas les Aristos*) has suggested similar caution with regard to the use of other Greco-Roman sources. In light of their warnings, therefore, let it be stated as a general basis of procedure that in seeking the milieu of the written gospel, we shall so far as is possible appeal to texts reasonably likely (that is, thought by modern commentators) to have originated within a century (three generations) before or after the period when the traditions about Jesus were formed and the gospels were written. In other words, we are looking for sources likely to come from between approximately 100 B.C.E. to 175 C.E.. In view of the more gradual nature of changes within Hellenism, we shall feel less pressured by that boundary in dealing with matters of Hellenistic culture than we shall in dealing with Judaism; in either case we shall be aware that many compositions from remote antiquity continued to exert their influence, as, indeed, they do to this day. Homer, Plato, Aristotle, and the Jewish Scriptures are obvious examples. Nonetheless, in the matter of literary milieu, the closer we can come to the center of our period, the more secure we shall feel.

BIBLIOGRAPHY

For an excellent general discussion of the issues in this chapter, see F. Gerald Downing, "A bas les Aristos. The Relevance of Higher Literature for the Understanding of the Earliest Christian Writings," *Novum Testamentum* 30.3 (1988): 212–30.

On the language question, see J. W. Voelz, "The Language of the New Testament," *ANRW* 25.2 (1984): 893–977. On the language of Mark, see the bibliography for Chapter 5.

On Hellenistic popular literature, see E. L. Bowie, "The Greek Novel," in P. E. Easterling and B. M. W. Knox, eds., *Cambridge History of Classical Literature*, vol. 1 (London: Cambridge University Press, 1985), 683–99; see further Tomas Hägg, *The Novel in Antiquity* (Berkely and Los Angeles: University of California Press, 1983); also B. P. Reardon, *The Form of Greek Romance* (Princeton: Princeton University Press, 1991). On the relationship of Hellenistic popular literature to the New Testament, see David E. Aune, *Greco-Roman Literature and the New Testament: Selected Forms and Genres* (Atlanta, Ga.: Scholars, 1988), especially 127–46 on the Greek novel, and literature there cited. Mary Ann Tolbert in *Sowing the Gospel: Mark's World in Literary Historical Perspective* (Minneapolis, Minn.: Fortress, 1989) has valuable observations about Mark's character as an example of popular literature (59–79), although her decision to regard this character as constitutive of Mark's genre is not, in my view, helpful. For extracts from some examples of Hellenistic popular literature, see the Appendix.

3

A Genre for Mark

I have yet to address directly the question asked earlier: What kind of text would Mark have been seen to be, either by its author or by its first readers and auditors? What, in short, was its genre?

Following the massive and influential work of Rudolf Bultmann, and for reasons that have perhaps more to do with theological conviction than scientific criticism, there has been until recently a curious reluctance among New Testament scholars to consider this question seriously, or to look to literature contemporary with the written gospel for parallels that will help us understand its genre. The written gospel, it has been claimed, is of its own kind: *sui generis*. It is unique, having no exact parallel in any literature. The gospels can be understood only as expansions of the primitive Christian preaching, or as having risen in response to a need for Christian lections to match the Jewish calendar in the liturgy. There are elements of truth in all these claims. There is evidence that early Christian literature was read in the Christian assembly (Col. 4:16; compare Philem 2). Yet there is hardly evidence that at this period it was read in sequence, or in connection with a "calendar" that is itself a matter of considerable scholarly uncertainty. Justin's description of the Roman liturgy in the mid-second century, when "the memoirs of the apostles or the writings of the prophets" were read "for as long as time permits," appears in any case to contradict such an idea (*Apology* 1.67). Certainly the gospels have connections with what we can discern or speculate of primitive Christian preaching, but then, it would be surprising if they did not. Certainly the gospels are in some sense unique. Yet from the viewpoint of literary genre, they have no more claim to be regarded as unique than have the writings of Paul, which we nonetheless seem willing to classify as letters (and even as "Hellenistic letters").

I have already said enough on the question of genre to indicate the unlikelihood that a document such as Mark, clearly popular in style, would be *sui generis*, or even that its author would be interested in stretching the boundaries of genre more than his subject absolutely required: genre is a tool of communication, and the evangelist is trying to communicate. No doubt Mark was well aware that the Word of the gospel was one that many would not or could not understand. That difficulty he would have regarded as arising from the nature of the Word itself, and from the state of the human heart (4:11–12). There is no evidence, however, that he chose to compound the problem by deliberately writing in ways that were hard to understand; on the contrary, as we have said, he wrote simply and popularly. Common sense suggests, therefore, that his intended genre would also be popular, and likely to be reasonably familiar to his readers.

My opening comments on Mark drew attention to its rather obvious Jewish connections, and it is natural therefore to begin by seeking contemporary literary parallels to it in Judaism. In fact, there are very few. Put at its simplest, Mark is a self-contained prose narrative centered upon the career and death of a single individual (that claim has itself been challenged, but I shall demonstrate the reasons for making it in a later chapter). I have already suggested that rabbinic literature is of limited relevance for our inquiry: granted the nature of Mark as we have described it, the rabbinic writings offer no parallels to it whatever. Only occasionally and incidentally can we infer from them any interest at all in details of the lives or deaths of the sages whom they quote. We do find in earlier Judaism some examples of self-contained narratives centered upon single persons, such as the books of Ruth, Esther, Jonah, Judith, and Tobit. But these generally focus upon one largely self-contained episode in the career of the protagonist and can hardly be said to portray her or his career and death. There are, again, elements of personal portrayal in the Old Testament story of Joshua, in the accounts of various leaders in Judges, and in narratives such as those concerning Elijah and Elisha, or David in the books of Samuel and Kings. Yet even with figures as powerful as these, it can never be said that their persons are in any sense central to the literature's purpose. Portrayal of the individual is always subordinate to an overall intention that extends well beyond that concern. The Old Testament prophetic books also contain several passages of biographical interest, yet here too it is finally God and God's purposes in the center of the stage.

The nearest Old Testament parallel to the shape of the written gospel is perhaps offered by the Pentateuch's account of Moses. The life of Moses, from birth to death, provides a framework for the books Exodus through Deuteronomy. Moses, like Jesus, is divinely appointed to office. The "books of Moses" combine teaching and event, as do the gospels. Like Jesus, Moses substantiates his message by performing miracles. Like Jesus, Moses is misunderstood even by his own. Much of the narrative centers upon a journey, moving from Egypt to the border of the Promised Land. The death of Moses itself is surrounded with mystery. Yet even here, it remains clear that the Pentateuch's central purpose is always to present God's deliverance of Israel, and God's commandments, not to portray Moses. Mark opens his gospel by speaking of "the beginning of the good news of Jesus Christ"; it is unimaginable that the Pentateuch could similarly have begun by speaking of the "good news of Moses." The Passover Seder to this day tells the entire Exodus story with scarcely a mention of Moses, in order (as the sages remind us) to make clear that it was God, not Moses, who delivered Israel. In so doing, the Seder is wholly faithful to the spirit of the Pentateuch.

Judaism more or less contemporary with Mark does however afford examples of autonomous narratives centering on single figures. *The Lives of the Prophets* (probably early first century C.E.) is a series of brief sketches giving details of the subject's birth and death, and some anecdotes—all of them, however, vastly shorter than the gospel. Philo of Alexandria, by contrast, wrote his rather lengthy treatise *On the Life of Moses*, combining chronological narrative adapted from the Scriptures with systematic analysis of exemplary qualities. In the case both of the *Lives*, however, and of *Moses*, while much of the content is Jewish, it is to Hellenistic literature, rather than Jewish, that we must look for the literary type. *The Lives of the Prophets* is most closely paralleled by a number of anonymous (sometimes pseudonymous) lives of Greek poets, generally fictional, frequently based on inferences from the subject's work, and often meant as an introduction to it (for example,

Life of Pindar). Philo's treatment is also, like the rest of his work, thoroughly Hellenistic. Here, however, is the clue: it is in fact in Hellenistic literary form that we discover the genre of the written gospel.

Reasonably well-educated Gentiles hearing or reading Mark at about the time of its composition would no doubt have found it somewhat strange. In particular, they would have been struck by its Jewishness and by its interest in eschatology. But they would hardly have thought they were encountering a new literary type. They would have perceived Mark as an example of what they would have called a "life" (βίος [transliterated *bios*]; Latin, *vita*), and what modern scholars loosely refer to as "Greco-Roman biography" (the noun *biographia* does not occur in extant literature until the fifth century C.E.). They would have made this identification between the written gospel and other "lives" for very good reasons; the following chapters will indicate what some of those reasons were.

BIBLIOGRAPHY

On the gospels as *sui generis*, see W. Schneemelcher in Edgar Hennecke and Wilhelm Schneemelcher, *New Testament Apocrypha* (Philadelphia: Westminster, 1963), 71–84. This view of the gospel is still virtually assumed in many studies: see, for example, Paul Achtemeier, *Mark* (Philadelphia: Fortress, 1986), 4–5, 42.

On the lectionary hypothesis applied to Mark, see Philip Carrington, *The Primitive Christian Calendar: A Study in the Making of the Markan Gospel* (Cambridge: Cambridge University Press, 1952); for discussion, see W. D. Davies, "Reflections on Archbishop Carrington's *The Primitive Christian Calendar*," in W. D. Davies and David Daube, eds., *The Background of the New Testament and its Eschatology. Studies in Honour of C. H. Dodd* (Cambridge: Cambridge University Press, 1956), 124–52.

On Old Testament models for the gospels, see Meredith Kline, "The Old Testament Origins of the Gospel Genre," *Westminster Theological Journal* 38 (1975): 1–27; Raymond Brown, "Jesus and Elijah," *Perspective* 12 (1971): 85–104; also Helmut Koester, *Ancient Christian Gospels: Their History and Development* (London: SCM; Philadelphia: Trinity, 1990), 27–29 and bibliography there cited. Vernon K. Robbins, *Jesus the Teacher: A Socio-Rhetorical Interpretation of Mark* (Philadelphia: Fortress, 1984) discusses perceptively the differences in emphasis and effect between the OT prophetic narratives and Mark (55–60).

On rabbinic parallels to the gospels (or rather, the lack of them), see Philip S. Alexander, "Rabbinic Biography and the Biography of Jesus: A Survey of the Evidence," in C. M. Tuckett, ed., *Synoptic Studies: The Ampleforth Conferences of 1982 and 1983* (Sheffield: JSOT, 1984); Jacob Neusner, *In Search of Talmudic Biography: The Problem of the Attributed Saying* (Atlanta, Ga.: Scholars, 1984).

On the gospels as Hellenistic "lives," see the bibliography for Chapter 5.

4

How to Show That Mark Is a Hellenistic "Life"

How do we establish that Mark's written gospel would have been perceived by contemporaries as a "life"? The problem is complicated by the fact that the ancients themselves did not delineate *bios/vita* as a genre. However, since (as we have observed) they also did not always take their "rules" too seriously, that may be less of a disadvantage than it at first appears. At least it relieves us of the temptation to try to read the written gospels against an alleged standard or prescription. The fact remains: whether the ancients delineated biography or not, they clearly did recognize some books as *bioi* or *vitae*, just as most moderns without a scrap of literary theory in their heads can perfectly well recognize a "western." How did they do it?

We have already noted that a genre is established by the dominant presence of a sufficient number of its particular themes or motifs. That does not mean that all the motifs of a genre must be present, or that motifs from other genres may not occur. Thus, even if it were true, as E. P. Sanders and Margaret Davies claim, that Mark differs from "a Hellenistic biography" in that "it does not begin with birth stories, and, if 16.8 is the original ending, it is quite without parallel," and moreover that in Mark "motifs are to be found which do not occur in Hellenistic literature, especially those associated with eschatological expectations" (*Studying the Synoptic Gospels*, 267, 270), still those factors alone would not mean that Mark would not have been recognized as a "life." (We do not, in fact, have to look far to find a Jew with fervent eschatological expectation who made use of an existing Hellenistic genre: Paul is the obvious example.)

David Aune (*New Testament*) and Richard Burridge (*What Are the Gospels?*) provide us with a more positive way of showing that the written gospel would have been perceived as a "life": they identify a sufficient number of the themes and motifs that are generic to Hellenistic "lives," and then show that from a literary point of view exactly the same generic themes and motifs dominate the written gospel, whatever its particular and individual deviations. This is the approach I intend broadly to follow with regard to Mark.

It is clear, of course, that many Hellenistic "lives" have perished, particularly among what would have been the less sophisticated examples. Nonetheless, a number have survived from the period under discussion: approximately 100 B.C.E. to 185 C.E. These include parts of the collections *On Distinguished Men* by Cornelius Nepos (c. 99–c. 24 B.C.E.) and Suetonius (born c. 69 C.E.); volumes of the *Parallel Lives* of Plutarch (before 50 C.E.–after 120 C.E.); individual works such as Lucian's *Demonax* (born c. 120); and anonymous popular works such as the *Life of Secundus the Philosopher* (late second century C.E.). Also relevant are texts such as the *Agricola* of Cornelius Tacitus (c. 56–c. 115 C.E.), which has many elements of a "life," though some critics deny that it was actually intended to be one, and Lucian's satirical *Passing of Peregrinus*, which, though epistolary in outline, contains a number of biographical motifs. Other "lives" that have frequently been compared with the gospels include Philostratus' *Life of Apollonius of Tyana*, Porphyry's *Life of Plotinus* and *Life of Pythagoras*, and Iamblichus's *Life of Pythagoras*. Their obvious similarities of subject (all are concerned with holy sages) make such comparisons tempting; but their lateness (all were written in the third or fourth century of the Christian era) makes the temptation one to be resisted. Indeed, the possibility that the gospels themselves have exercised literary influence here cannot be ruled out.

The attentive reader may suggest at this point that the process I propose is circular. I have identified certain texts as Hellenistic "lives," presumably on the basis of their characteristics, and now plan to examine them to see what are the characteristics of Hellenistic "lives."

It should be noted, however, that it was the ancients themselves who, for the most part, identified the texts I shall cite as *bioi* or *vitae*. What I propose is simply to point to the considerable cluster of generic elements these texts have in common, and to show that this same cluster is also present in Mark. On that basis I shall argue that they and Mark are properly regarded as examples of a single genre. That is the real question.

What particular elements in a text should be examined as indicators of genre? There can be no definitive answer. Various critics suggest various schemes, usually trying for a balance between internal and external characteristics: thus we have form, content, and function suggested by B. H. M. G. M. Standaert (*L'Evangile selon Marc*, 619–26) and David Aune (*New Testament*, 46–63), form and material suggested by Robert Guelich ("The Gospel Genre," 213), and "internal features" and "external features" by Richard Burridge (*What Are the Gospels?* 111). We cannot here enter into the complex question of the relationship between form and content; indeed, the distinction implies a dichotomy that cannot finally be maintained. What is important, and what all informed schemes have in common (Burridge's being by far the clearest) is that we must identify enough characteristic elements to establish whether or not in any given work they dominate: only so may we avoid the mistake of confusing subordinate motifs (such as the love affair in *High Noon*) with genre. In the following chapters, therefore, we shall consider certain elements, reflecting both "form" and "content," but not attempting to make a formal distinction between those two aspects. The selection of features is basically adapted from that drawn up by Burridge (*What Are the Gospels?* 112–27):

1. title
2. opening features
3. subject
   (a) establishing who the subject is
   (b) establishing the type of subject
4. geographic setting
5. arrangement of material
   (a) structure
   (b) allocation of space
6. characterization
7. sources and units of composition

8. common motifs
   (a) origins, birth, education, and early exploits of the hero
   (b) deeds of the hero
   (c) death of the hero
9. written style
10. length
11. function

BIBLIOGRAPHY

On Hellenistic "lives" generally, see Arnoldo Momigliano, *The Development of Greek Biography* (Cambridge: Harvard, 1971). For discussions of the gospels as Hellenistic "lives," see David E. Aune, *The New Testament in Its Literary Environment* (Philadelphia: Westminster, 1987), 17–76; also David E. Aune, ed., *Greco-Roman Literature and The New Testament: Selected Forms and Genres* (Atlanta, Ga.: Scholars, 1988), 107–26. Richard A. Burridge, *What Are the Gospels: A Comparison with Graeco-Roman Biography*. SNTS Monograph Series 70 (Cambridge: Cambridge University Press, 1992) offers the most detailed examination of the problem and its solution, and is, in my view, definitive. Among earlier writers, C. W. Votaw, "The Gospels and Contemporary Biographies," *American Journal of Theology* 19 (1915): 45–73 and 217–49 should be noted. Post-Bultmann, the insightful comments of Stephen Neill in his Firth Lectures for 1962 deserve mention (*The Interpretation of The New Testament 1862–1961: The Firth Lectures, 1962* [London: Oxford University Press, 1966], 259–62), but credit for the first full-length attempts to establish the biographical nature of the gospels must go to Charles H. Talbert (*What Is a Gospel? The Genre of the Canonical Gospels* [Philadelphia: Fortress, 1977]) and P. L. Shuler (*A Genre for the Gospels: The Biographical Character of Matthew* [Philadelphia: Fortress, 1982]). Unfortunately, neither study is without serious problems: for devastating criticism, see D. E. Aune, "The "Problem of the Genre of the Gospels: A Critique of C. H. Talbert's *What is a Gospel?"* in R. T. France and D. Wenham, eds., *Gospel Perspectives: Studies of History and Tradition in the Four Gospels* 2 (Sheffield: JSOT, 1981), and R. A. Burridge's review of P. L. Shuler's *A Genre for the Gospels* in *Anvil* 2.2 (1985): 179–80.

As regards the genre of Mark in particular, Vernon K. Robbins, (*Jesus the Teacher: A Socio-Rhetorical Interpretation of Mark* [Fortress: Philadelphia, 1984]) notes the gospel's "significant parallels to contemporary Greco-Roman biographies" (4), but since he offers little to support this observation save brief references to Votaw and Talbert, his discussion can hardly be regarded as satisfactory. Mary Ann Tolbert (*Sowing the Gospel: Mark's World in Literary Historical Perspective* [Minneapolis, Minn.: Fortress, 1989]) has some useful initial remarks about genre, but then confuses the question by attempting to establish Mark's genre almost solely on the basis of its character as popular literature (48–79).

On Cornelius Nepos, see Nicholas Horsfall, *Cornelius Nepos: A Selection Including the Lives of Cato and Atticus* (Oxford: Clarendon, 1989); on Plutarch, see Christopher B. R. Pelling, "Plutarch's Method of Work in the Roman Lives," *JHS* 99 (1979): 74–96; on *The Life of Secundus the Philosopher*, see Ben Edwin Perry, *Secundus the Silent Philosopher* (New York: American Philological Association, 1964); on Suetonius, see Andrew Wallace-Hadrill, *Suetonius: The Scholar and His Caesars* (New Haven: Yale, 1983); on Tacitus, see Ronald Martin, *Tacitus* (Berkeley and Los Angeles: University of California Press, 1981); on the "lives" of late antiquity, see the elegantly written study by Patricia Cox, *Biography in Late Antiquity: A Quest for the Holy Man* (Berkeley and Los Angeles: University of California Press, 1983).

5

Mark as a Hellenistic "Life"

Title

Titles, as we have observed, often indicate genre. The problem with the titles of ancient texts is that we are not always sure that they are original. Even when not original, however, they at least tell us how the work was perceived by early readers; from our point of view that is almost as valuable as the author's own opinion. Some of the writings with which we are comparing Mark are directly called "lives," such as Lucian's *Life of Demonax* (ΔΗΜΩΝΑΚΤΟΣ ΒΙΟΣ), and the *Life of Secundus the Philosopher* (*ΒΙΟΣ ΣΕΚΟΥΝΔΟΥ ΦΙΛΟΣΟΦΟΥ*). Others are simply called by the name of their subject, like Plutarch's *Demosthenes* and *Cicero*. Plutarch himself, however, refers to his entire collection as *parallel lives* (τῶν παραλλήλων βίων; see *Demosthenes* 3.1). Similarly, Suetonius's magnum opus on the twelve Caesars appeared under the general title *On the Life of the Caesars* (*De vita Caesarum*).

The title of the written gospel, [*Gospel*] *according to Mark* ([ΕΥΑΓΓΕΛΙΟΝ] ΚΑΤΑ ΜΑΡΚΟΝ [transliterated (*euaggelion*) *KATA MARKON*]) is certainly not original, though it may be as old as the beginning of the second century. The shorter form (preserved by codices Sinaiticus and Vaticanus) implies the longer (preserved by codex Alexandrinus); in either case *kata* ("according to") serves to indicate that the gospel is perceived as an example. But an example of what? Disappointingly for our hypothesis, it is not genre that is here under consideration at all, but content. In the New Testament, *euaggelion* refers to the proclamation of God's saving work: this is clear in Paul (who uses the word more than forty times) and in Mark (1:14; compare 8:35, 10:29, 13:10, 14:9). *Euaggelion* speaks of a theological claim, not a literary type. So used, it occurs always in the singular, for there can only be one gospel (Gal. 1:7–9). This is the usage reflected by the written gospel's title. Paul too spoke of his "gospel" (that is, the gospel according to Paul), but used quite a different genre (Rom. 2:16). Not until the middle of the second century do we find the word *euaggelion* used as the name for a particular kind of text, and hence appearing in the plural (*euaggelia*, "gospels") (2 *Clem.* 8:5, Justin, *Apology* 1.66.3, *Dial.* 10.2, 100.1). Among Gnostics, on the other hand, *euaggelion* appears in time to have become a catch-all for any kind of "Jesus literature," as witness the diverse group of writings (and genres) to which it is applied.

Opening Features

Epic poets like Homer, as Horace observed, plunged at once into the midst of the action (*festinat et in medias res*) (*Art of Poetry* 148). "A to B, greeting," as every student of Paul's writings knows, was the usual way to begin a Hellenistic letter. Such openings signal to the reader the kind of text that is to follow. With a "life," as with a "history," there would sometimes (but by no means always) be a prologue or preface indicating the author's reasons for writing, and intentions. Then the narrative proper would begin, with the subject's name occurring within the first few words. Thus Plutarch opens his *Demosthenes* with a delightful, self-deprecating preface, full of good humor and common sense, in which he apologizes for his inability to make any useful comparison between the oratory of Demosthenes and that of Cicero, owing to his inadequate appreciation of Latin, but explains why, nonetheless, he considers it appropriate to recount their lives in parallel. Then at last he begins a new period: "Demosthenes the father of Demosthenes. . . ," the formal statement of name indicating at once that the writer has concluded his prefatory remarks and is now moving to his task (*Demosthenes* 4.1). Using the same structure, if less gracefully, Philo explains to us why he undertakes "to write the life of Moses," and then begins with what, as he pedantically points out, "is necessarily the right place to begin: Moses was by race a Chaldean, but was born in Egypt. . . ," and so on (*Moses* 1–5).

On other occasions Plutarch reverses the process, beginning with the name of his subject: "It is the life of Alexander the king, and of Caesar . . . that we are writing" [Τὸν Ἀλεξάνδρου τοῦ βασιλέως βίον καὶ τοῦ καίσαρος . . . γράφοντες] (*Alexander* 1.1), and then makes some prefatory remarks (1.1–3). On other occasions, as in *Cicero* and *Cato the Younger*, he simply begins with the name of his subject: "They say that Helvia, the mother of Cicero . . ." [Κικέρονος δε τὴν μὲν μητέρα λέγουσιν Ἑλβίαν . . .] (*Cicero* 1.1), and proceeds straight to his matter. (The inflected nature of Greek allows, indeed, a prominence for this first mention of the name that is hard to render into English.) In connection with this last type, we should probably refer to what Lucian calls a "virtual preface" (προοίμιον δυνάμει; *How to Write History* 23, 52). Some subjects, he suggests, do not need preliminary explanation. In these cases, good writers will simply ensure that their opening statements are understandable, and that they prepare their audience for what is to come.

Often a writer will on this first occasion give the subject's name more fully than elsewhere in the work, perhaps adding titles (compare Plutarch's "Alexander the king" above). At the conclusion of Tacitus's lengthy prologue to the *Agricola*, he begins his account proper with the words, "Gnaeus Iulius Agricola" (*Agricola* 4.1.)—the only time in the whole work where he refers thus formally to his subject by *praenomen*, *nomen*, and *cognomen*. The *Life of Secundus,* by contrast, offers another "virtual preface," having no introductory remarks of any kind: but it does begin with the hero's name, and the one title that is significant. "Secundus was a philosopher" (Σεκοῦνδος ἐγένετο φιλόσοφος) (*Secundus* [Perry] 68 1).

Finally we should note that many of these beginnings—including some that plunge at once into their subject, and some that approach it more indirectly by way of prologue—are strong rhetorically. They invite or provoke the audience's attention. They persuade us to sympathize with the narrator or the subject. At the very beginning of this study we noted that authors at this period would normally expect their work to be experienced by being heard as it was read aloud. Let us note how effective many of the openings we have considered would be as rhetoric: gently winning our sympathy and interest, as do *Demosthenes* and the *Agricola*, or taking us at once into the midst of the story, as does *Cicero*.

Mark's opening,

> The beginning of the good news of Jesus Christ, [son of God], as it is written in Isaiah the prophet . . . ! (1:1–2) (RSV, punctuation altered: see Chapter 9)

falls well within the boundaries we have described. Its immediate plunge into action naturally reminds us of Lucian's "virtual preface." It includes Jewish and eschatological motifs that would no doubt have seemed strange (and perhaps therefore interesting) to average Hellenistic listeners, but it also does exactly what they would have expected in a "life": it prominently and formally names the subject, and gives his title. The additional title "son of God," even if not original, suggests at least the work of an early copyist who also thought of Mark as a "life," knew what was proper for the beginning of such a work, and made an appropriate addition.

Mark's opening is, moreover, particularly striking as rhetoric. Like the beginning of any effective speech, it is designed to get the audience's attention. It bids us look to the matters to be handled, placing them vividly in the setting of lengthy antiquity and divine promise. This is not, of course, the rhetoric of civic or intellectual life, which seeks to gain the audience's favor by persuasion, charm, or reason. We are a long way from Plutarch (or even Luke). This is religous rhetoric, described by George Kennedy as "a form of 'sacred language' characterized by assertion and absolute claims of authoritative truth without evidence or logical argument" (*New Testament Interpretation*, 104). Such rhetoric was perfectly familiar to those who knew either the Hebrew prophets or the literature of classical antiquity, or even to those who had merely heard Cynics and others declaiming those "street corner invocations to Virtue" which Lucian so detested (*Peregrinus* 3; compare 4).

Subject

Establishing Who Is the Subject

Common sense, faced with Lucian's *Demonax* or a written gospel, is inclined to reply somewhat testily to the question Who is the subject? by observing that in each case the answer is perfectly obvious. Common sense is probably right. Nonetheless, the question has been posed, at least as regards the gospels. Robert Guelich asks: "In short, does the evangelist view his task to write a 'biography' or to set forth the Christian message about what God was doing in and through Jesus Messiah? Is the ultimate focus not on God rather than on Jesus . . . ?" ("The Gospel Genre," 191). In a similar vein, David P. Moessner suggests that "the overriding question that emerges both within this narrative world and for the reader is *not*, 'What sort of person is this,?' but rather, 'Who is this person in the light of God's dealings with Israel?'" ("And Once Again," 76). Admittedly, a great deal is blurred in these passages by the use of the words "ultimate" and "overriding." Still, it is obvious where Guelich and Moessner are heading. The same kind of thinking is manifested by Herman C. Waetjan, who says that Mark "is not to be identified with the literary genre of biography. Its subject matter is neither mimetic nor historical but ideological" (*Reordering of Power*, 1–2).

Richard Burridge offers a simple but convincing literary tool to identify the subjects of selected Hellenistic "lives." In a group of five that includes Tacitus's *Agricola*, Plutarch's *Cato the Younger* and Lucian's *Demonax*, he notes in terms of percentage the proportion of occasions throughout the text in which the names of significant characters occur in the nominative, observing that "*bios* literature is characterized by a strong concentration and focus on *one* person, and this is reflected even in the verbal syntax" (*What Are the Gospels?*, 163). By way of control he compares other texts, such as the *Iliad* and the *Odyssey*, which are not "lives," but describe obviously striking characters. It is not necessary here to repeat the details of Burridge's analysis. Suffice it to say that in the *Iliad* and the *Odyssey* the highest score registered by any single character is 4.8 percent by Odysseus in the *Iliad*, but there is quite a range of other characters with comparable scores. By contrast, in the *Agricola* Agricola scores over 18 percent, in *Cato the Younger* Cato scores 14.9 percent, in *Demonax* Demonax scores 33.6 percent; no other character in any of the three "lives" registers a score that is anywhere near that of the main subject. Mark, similarly analyzed, produces a score of 24.4 percent for Jesus, with all other characters virtually nowhere. In other words, regardless of what theological subtleties may have been lurking in the back of Mark's mind, his immediate concern appears to have been to write about Jesus. "The primary interest of our earliest evangelist is in the significance of the person of Christ" (R. H. Lightfoot, *History and Interpretation of the Gospels*, 61). As a result Mark has produced a text which in its verbal emphasis on Jesus corresponds exactly to the emphasis we expect to find in a Hellenistic "life."

One qualification should be noted: the foregoing combines results from texts that were counted by hand (the *Agricola*, *Demonax*, and Mark) and *Cato*, which was analysed by computer. The general tendency of the figures is such, however, as to render it unlikely that a handcount of *Cato*, when there is time for one, will significantly alter the result. Indeed, if anything, handcounting seems likely to strengthen the majority in favor of the hero.

Type of Subject

Within our period, three kinds of persons generally seem to have been regarded as the proper subjects of Hellenistic "lives": public figures such as statesmen, generals, and monarchs; literary figures such as orators or poets; and philosophers and sages. Examples of this last category would be the heroes of *Demonax* and the *Life of Secundus*. In Mark's presentation of Jesus, while there are overtones of royalty, it is clearly the sage with whom he has most in common. Jesus, like Demonax and Secundus, answers the questions of those who come to him, confounds friends and enemies alike with his wisdom, witnesses fearlessly to the truth before authorities, and is willing to suffer for his witness. More than one critic has noted similarities of form and even content between the brief units in which much of the teaching of Jesus is presented, and the *chreiae* that typically enshrined traditions about Diogenes the Cynic and other teachers. Yet there are also differences. Secundus and Demonax do not work miracles, and Jesus is not merely a sage and teacher: he is also God's beloved Son (1:1, 1:11, 9:7, 12:6). In this matter, Mark was constrained by the tradition he had received. Those elements in the church's memory of Jesus that spoke of him as exorcist, miracle worker, and Son of God, could not be avoided.

In being so constrained, Mark perhaps took the "life" genre in a new direction, stretching its boundaries in new ways. Yet if in this respect his writing was without precedent, it was certainly not without successors. Most obvious are the other gospels. From the viewpoint of literary history, however, equally interesting are various "lives" of semidivine sages written from the third to the fourth century: notably, Philostratus's romantic *Life of Apollonius of Tyana*, Porphyry's *Life of Pythagoras*, and Iamblichus's *Life of Pythagoras*. All describe holy men who are models of virtue and wisdom. Their births are accompanied by miracle, they perform miracles, they teach, they are compassionate toward those around them, and they suffer misunderstanding from their enemies and their friends. We have said that these writings are too late to be regarded as evidence of Mark's literary milieu; but it is quite another matter to regard them as in some measure his literary successors. In this connection, we should also refer again to Lucian, whose semibiographical *Passing of Peregrinus* (written shortly after 165) is an amusing satire on a man who posed as sage, miracle worker, and prophet, and who was regarded by Lucian as a complete charlatan. Lucian already appears to be aware of the elements that go to make up the later "lives of divine sages," and spoofs them all. Peregrinus suffers persecution at the hands of authority for his pretended frankness of speech, demands extreme asceticism from all, is regarded by his disciples as a miraculous healer, and even enjoys a spurious "resurrection" after death (*Peregrinus* 18, 19, 28, 40). Such jokes depend upon recognition.

There are, of course, important differences between the gospels and these later writings. We note three differences in particular: first, the gospels' interest in eschatology; second, the gospels' Jewishness (by contrast, Philostratus's Apollonius regards even the land of Israel as polluted because of the acts and sufferings of its people [5.27]: a fascinating mirror image of Israel's self-perception); and third, the gospels' comparative *lack* of interest in asceticism, particularly as regards diet (thus, in contrast to Philostratus's Apollonius, who claims that eating meat is unclean, the disciples of Jesus are told bluntly that "there is nothing outside a person that by going in can defile" [see *Life of Apollonius* 1.8; compare Porphyry's *Life of Pythagoras* 7.34–35; contrast Mark 7:15]). Granted these differences, it remains that any complete account of the evolution of later "lives of divine sages" probably ought to find a place for Mark and his fellow evangelists.

Setting

The question of setting appropriately follows that of subject, for it is the subject who always determines the setting of a "life." With Philo's *Abraham* and *Moses*, or Plutarch's *Alexander*, we travel far, because the protagonist travels. With *Secundus* we are first in lodgings in his unnamed hometown, then in the house of his mother, then before the emperor Hadrian in Athens, because that is where Secundus goes. With Lucian's *Demonax*, we hardly move at all, because Demonax does not.

The one rule for the setting of a "Life" is that the subject is center stage. Even if, for a while, this appears not to be the case—as with Philo's discussion of religious laws (*On the Life of Moses* 2.17–65) or details of the tabernacle (2.77–140), or Secundus's answers to Hadrian's questions (Perry 78.11–90.12)—the digression is only apparent. The laws and the details of the tabernacle are being discussed because Moses established them. It is Secundus himself who is giving the answers.

All this, again, is the pattern of Mark. First we are in and around Galilee, then we are on the road to Jerusalem, and finally we are in and around Jerusalem, *with Jesus*. The narrative appears to break this rule only rarely. Briefly, when we focus on the wilderness and the Baptist (1:2–8); again, when we travel without Jesus to the court of King Herod (6:14–29); again, when Judas offers to betray Jesus (14:10–11); and again, when Peter disowns Jesus (14:54, 66–72). In each case the diversion of attention is only apparent, not real. Jesus is always the true center of concern. Thus, the portrayal of the Baptist is, as we shall see, simply a way of pointing to the significance of Jesus, and, from the viewpoint of literary structure plays exactly the same role in the gospel that the account of the dreams of Olympias and Philip plays in Plutarch's *Life of Alexander* (2.1–2). The novella of discussion and death at King Herod's court similarly prepares us for the conversation at Caesarea Philippi and for the Passion of Jesus (6:14–16 [compare 8:27–30]; 6:27–29 [compare 9:9–13]). The account of Judas's treachery inaugurates Jesus' Passion—betrayed by one of his own. The story of Peter's denial serves above all to emphasize Jesus' command of events—for Jesus foretold it (14:29–30).

Arrangement of Material

Structure

Friedrich Leo, in his monumental study *Die griechisch-römanische Biographie nach ihrer literarischen Form* (1901), distinguished two structural forms in Greco-Roman biography: the Plutarchian, and the Suetonian. The Plutarchian involved a broadly chronological narrative throughout and allowed character to emerge through actions; the Suetonian largely abandoned chronological narrative and, between accounts of the subject's birth and death, provided systematically arranged accounts of qualities and accomplishments. The former was suitable for statesmen, generals, and philosophers; the latter for writers and artists. Like most such broad categorizations, Leo's distinction, though suggestive, must be used with caution, particularly since it does not allow for mixed examples, of which there are many. Among the ancient precedents, Xenophon's *Agesilaus* (written c. 360–356 B.C.E.) appears to combine both methods: the first half dealing with the hero's life chronologically, the second giving an account of his virtues. Among writers closer in time to Mark, Cornelius Nepos follows the same arrangement in his *Atticus*, and so does Philo in his *Moses*, as Philo himself is at pains to point out (*On the Life of Moses* 1.334, 2.1–7). Even Plutarch does not hesitate in his *Life of Cato the Elder* to introduce a collection of Cato's remarks with the words, "I shall now relate a few of Cato's memorable sayings," concluding equally bluntly, "These are some examples of his memorable sayings" (*Cato*, 7, 9; compare Mark 4:2, 4:33–34). Perhaps the most that safely can be said is that the majority of Greco-Roman "lives" are set within a superficially chronological framework, moving from birth to death. Some, like Tacitus's *Agricola* also follow a largely chronological sequence in between, while others, like Lucian's *Demonax*, are just loose strings of anecdotes. In such company, Mark's narrative certainly needs no apology or explanation.

On the one hand, Mark has a basically chronological organization from Baptism to Passion (via the Ministry in Galilee, the Crisis at Caesarea Philippi, and the Journey to Jerusalem). Within that outline, there are further ordering references, both backward and forward—backward, in references at various points in the narrative to earlier events, such as Jesus' Baptism as the source of his authority (11:27–33) and Jesus' previous actions as reasons why his disciples should understand him (8:16–21), and forward, in repeated anticipations of the coming Passion and resurrection (2:20, 3:6, 8:31, 9:9, 9:31, 10:30–34, 10:45, 14:7–8). Once there is a flashback, in the account of intrigue surrounding the arrest, imprisonment, and subsequent death of John the Baptist (6:16–29; compare 1:14).[1] In addition, there are further refer-

1. Mark may here be something of an innovator. Of course flashback is common enough in Homer—indeed it is, as we shall have occasion to notice, a device indispensable in oral poetic narrative (see Chapter 7). In *prose* narrative, however, B. P. Reardon is unaware of any use of this device before that of Heliodorus in his romantic *Ethiopica* (*Form of Greek Romance*, 40). Since the *Ethiopica* was composed two or possibly three hundred years later than Mark, this puts the evangelist well in the vanguard.

ences forward to events not actually narrated by Mark but within the memory of his audience, such as the persecution of Christians and the coming Fall of Jerusalem (13:9–23), and to the final presence of the Son of man in the unknown future (13:26, 14:62). On the other hand, as is evident when we examine the "lives" to which I have just referred, Mark's insertion of other material by topic, such as the collection of parables (4:1–34), or a series of mighty acts (4:35–5:43) would not have seemed to a contemporary auditor in the least strange or reprehensible.

Allocation of Space

Martin Kähler's often-quoted observation that the gospels are "passion narratives with extended introductions" (*Der sogenannte historische Jesus*, 80) has perhaps been taken more seriously than it deserves, or than he intended. (Critics generally cite Kähler in this connection as if he had spoken with a perfectly straight face: it is worth recalling that he offered his description "somewhat provocatively" ["*etwas herausfordernd*," 80]). Be that as it may, the argument that the written gospels cannot be biographies because of the disproportionate amount of attention they devote to Jesus' death has often been used. Actually, as Richard Burridge points out, it is quite usual for Hellenistic biographers to give what we might regard as a disproportionate amount of space to particular items (*What Are the Gospels?* 164–67). Plutarch devotes 12.8 percent of *Demosthenes* to the death of Demosthenes, and 17.3 percent of *Cato the Younger* to the death of Cato. Tacitus devotes 26 percent of the *Agricola* to the Battle of Mons Graupius. *Secundus* devotes 14.3 percent to the "Fateful Reunion," 26.4 percent to the "Test" before Hadrian, and 58.3 percent to Secundus's teaching. With these figures, Mark's 18.4 percent for the Passion of Jesus, and 20.2 percent for his teaching, compare quite reasonably.

Hellenistic biographers seem, in fact, to have thought it perfectly normal and acceptable to devote "disproportionate" amounts of space to whatever in their subjects' "lives" or words they regarded as especially important or interesting. In this, apparently, they reflected the critical views of their time. Lucian expects good writers to "run quickly over small and less essential things, while giving adequate treatment to matters of importance." By contrast, something that sounds remarkably like the "balanced" treatment looked for by modern critics he regards as a sign of poor taste: ". . . indeed, a great deal should be omitted. When you feast your friends and all is ready you do not for that reason in the middle of all your pastries, fowl, oysters, wild boars, hare, and choice fish cutlets serve up salt fish and pease porridge because that, too, is at hand—you will ignore the humbler fare" (*How to Write History* 56, trans. K. Kilburn). Lucian was, of course, speaking of history, but we may reasonably suppose (and, indeed, from the material available to us it is evident) that his opinion represents an attitude prevailing equally in regard to related genres such as the "life."

We should further note that brief "lives" describing the deaths of notable persons seem to have been particularly fashionable toward the latter part of the first century C.E. Gaius Fannius wrote about the deaths of famous men executed or banished under Nero (Pliny, *Letters* 5.5) and Gnaeus Octavius Titinius Capito's *Passing of Famous Men* (*exitus inlustrium virorum*) commemorated the deaths of republican martyrs (Pliny, *Letters* 8.12). The death of Socrates continued to have a powerful effect on the imagination of Christian and pagan alike. In this connection, undoubtedly, we should also remember Lucian's satirical *Passing of Peregrinus*, which presents an obvious spoof of the theme.

Characterization

There is an often-quoted passage in Plutarch's *Alexander*, in which he echoes (consciously or not) a thought that had earlier been set down by Cornelius Nepos, and apologizes for the amount of material he has been obliged to omit:

> For it is not Histories that we are writing, but Lives; and in the most illustrious deeds there is not always a manifestation of virtue or vice, nay, a slight thing like a phrase or a jest often makes a greater revelation of character than battles where thousands fall, or the greatest armaments, or seiges of cities. . . . so I must be permitted to devote myself rather to the signs of the soul in men, and by means of these to portray the life of each, leaving to others the description of their great contests. (Plutarch, *Alexander* 1:2–3, trans. Bernadotte Perrin; compare Cornelius Nepos, *Pelop.* 1.1)

As Patricia Cox has expressed it, "The biographer's task was to capture the gesture which laid bare the soul" (*Biography in Late Antiquity*, xi).

Plutarch's talk of "virtue" (ἀρετή) and "vice" (κακία) does not mean that he, or other writers of "lives," merely moralize; it does mean that we are seldom in much doubt as to whether we are to admire and emulate the subjects of their work or not. Thus Lucian first tells us his high opinion of Demonax's virtues (5–10), then gives us a series of episodes and anecdotes that illustrate his character. Even with *Secundus*, while we find no expression of outright authorial opinion, we can scarcely be unaware that we are to perceive a man who was, despite his early folly, both wise and brave.

What we do not find in Hellenistic "lives" is much analysis of character or motive in the modern sense. Certainly Plutarch is aware of the contradictions in, say, a Cicero or a Demosthenes, contrasting the former's "immoderate boasting" and "intemperate desire for fame" with his genuine "contempt for wealth and . . . his humanity and goodness" (*Demosthenes and Cicero* 2.1, 3.3). Yet Plutarch still seems to see these contrasts in terms of being vicious or virtuous. He gives little hint of any awareness of contradictions and complexities within the personality itself. He records episodes from his protagonist's childhood or youth, but in general this is not to show how the protagonist developed or changed; rather, what such episodes most often demonstrate is that for those with eyes to see it was evident from the beginning what this child would become (for example, *Alexander* 4.4–10.4; *Cicero* 2.2–4). Similarly, when Suetonius tells us how the emperor Nero began his reign with "acts, some of which are beyond criticism, while others are even deserving of no slight praise" (*Nero* 19.3; compare 8–19), he does not suggest that the emperor changed, but rather that his true colors, at first hidden, were finally revealed.

> . . . at first his acts of wantonness, lust, extravagance, avarice and cruelty were gradual and secret, and might be considered as follies of youth, yet even then their nature was such that no one doubted that they were defects of his character (*naturae illa vitia*), and not of his time of life. . . . Little by little, however, as his vices grew stronger, he dropped jesting and secrecy, and with no attempt at disguise openly broke out into crime. (*Nero* 26.1, 27.1, trans. J. C. Rolfe)

Such characteristics as these have led Christopher Gill to say of Plutarch, in particular, that he is primarily interested in "character," as opposed to modern biographers who are more concerned with "personality." By "character," Gill means

> (i) placing people in a determinate ethical framework and (ii) treating them as psychological and moral "agents," that is, as the originators of intentional actions for which they are normally held responsible and which are treated as indexes of goodness or badness. ("The Character-Personality Distinction," 2)

By "personality," Gill means

> a response . . . that is empathetic rather than moral: that is, with the desire to identify oneself with another person, to "get inside her skin," rather than appraise her "from the outside" . . . concern with the person as a unique individual . . . rather than as the bearer of character-traits which are assessed by reference to general moral norms . . . a perspective in which the person is seen as psychologically passive; that is, as someone whose nature and behaviour are determined by forces which fall outside her control as an agent and perhaps outside her consciousness as well. (2)

This insight is valuable, not merely in regard to Plutarch, but in regard to Hellenistic "lives" generally. It cannot, of course, have been that Hellenistic biographers were unaware that personal history and events over which one has no control can lead to surprising aspects of individual personality and behavior: how could those who had grown up with Greek tragedy *not* have thought of this? Plutarch at times shows such awareness, as in his description of Coriolanus's relationship to his mother, and even his note about Lysander's upbringing (*Coriolanus*, 4.3–4; *Lysander*, 2.4). Yet even in Plutarch the analysis is never profound, and neither of him nor of any other Hellenistic biographer can we claim that this is a center of their interest. What seems to have interested them seems rather to have been behavior that was *typical*, the sort of thing one would expect of such a person—and in that, more perhaps than anything, they differed from biographers in the nineteenth and twentieth centuries. This leads to a result that Christopher Pelling has described perfectly in reference to Plutarch's Antony:

> Plutarch individuates his personalities; he has a rich and differentiated vocabulary for describing traits; but it remains true that he, like most or all ancient writers, has an extremely *integrated* conception of character, and that his figures are consequently individual in a way which we find oddly limited. The differing elements of a character are regularly brought into some sort of relationship with one another, reconciled: not exactly unified, for a character cannot be described with a

> single word or category, and is not a stereotype; but one element at least goes closely with another, and each element predicts the next. ("Childhood and Personality," 235)

Thus, as Pelling points out, Plutarch's Antony has his simplicity (ἁπλότης), which leaves him vulnerable to flatterers and more powerful personalities such as Curio, Fulvia, and Cleopatra; and that quality helps explain why he is at times so passive. His simplicity in turn goes well with his soldierliness, and with the sense of fun he shares with his men, and later with Cleopatra: thus the same qualities both build him up and ruin him. Soldierliness goes well with the nobility he shows, for example, at Philippi, when he honors the fallen Brutus (22.6–7); and that same nobility goes with his capacity to care for Roman values and duty, and so to feel deeply his final disgrace (67, 76). "It all fits together very tightly: not as a stereotype, for these are distinct traits; but they are closely *neighbouring* traits, and we are not surprised that Antony shows them all" ("Childhood and Personality," 235–36).

In view of this, it is perhaps not surprising that there was some tendency among Hellenistic biographers to characterize subjects by *type*, and we find sets of "lives" based on type as their theme. Cornelius Nepos's *On Distinguished Men* (which included foreigners as well as Romans) contained the categories of generals (which survives), historians, kings, poets, and possibly also orators. Suetonius's work by the same title dealt with Roman men of letters (of which, alas, only the grammarians and the rhetors have survived—we should have preferred the poets), and his *Lives of the Caesars* dealt with emperors. Plutarch's "parallel lives" dealt with Greek and Roman statesmen. It is broadly in line with this that in many respects a characterization that is typical rather than individual seems to emerge even in "lives" that are not parts of sets: Tacitus gives us Agricola the model of the good soldier, and Lucian in *Demonax* and the author of *Secundus* both, in their own ways, give us models of the philosopher.

Into this pattern Mark fits well. Mark, too, as has often been observed, shows no interest in character development or personality as understood by moderns, and makes not the slightest attempt to view his subject from inside, or get "under his skin." As R. H. Lightfoot put it, "we hardly ever see [Jesus], except as a teacher or mighty worker or engaged in controversy; above all, we are not admitted to a knowledge of his inner life" (*History and Interpretation*, 99). Mark's Jesus is, in Gill's sense, definitely a "character" rather than a "personality." He is fully responsible for all that he does, and is never, save physically at the Passion, passive or acted upon. Even throughout the Passion he remains sovereign, the Son of God, as is finally witnessed by the centurion's confession (15:39).

Mark's Jesus is, moreover an integrated personality in Pelling's sense: that is to say, he is clearly defined by certain key characteristics that go naturally together. He is, above all, one "having authority" (1:22; compare 1:27 and 11:27–33)—the *strong* Son of God. This authority is manifest in his dealing with opponents (2:12, 3:1–6), demons (1:23–27, 9:25–27), and disciples (1:16–20). The same authority that will confront the demoniac at Nazareth will also twice refuse to have itself directed by Peter (1:36–39, 8:32–33), and will remain in command even amid the humiliation of arrest and trial (14:48–49, 14:62). Alongside the authority goes great passion: anger at hardness of heart (3:1–6), the harsh "Get behind me, Satan!" of Caesarea Philippi (8:33), the angry confrontations of the final visit to Jerusalem (11:12–14, 27–33), the agony in the Garden of Gethsemane (14:33–36), and the final cry of desolation when, alone and forsaken, he dies (like Socrates) at the hands of the unjust. Alongside the Passion goes great *com*passion, displayed in scene after scene, not only with the sick and the weak, but also with his disciples at their stupidest and most recalcitrant (10:35–45). Authority, passion, and compassion: entirely conventional characteristics for one who is Messiah and Son of God. In other words, Jesus is entirely the sort of person one would expect, which is in turn precisely what we should expect in a Hellenistic "life."

Mark's techniques for presenting his subject are, like his understanding of it, conventional. He expresses directly his own authorial opinion of his hero by his opening words, "Christ, [Son of God]," and by at once applying to him the words of God's promise (1:1–3). John the Baptist speaks of Jesus, Mark tells us, "as it is written, . . . the voice of one crying, Prepare the way of the Lord." Thereafter, who Jesus is and what he is emerges through his words and actions. To this end are employed sayings (4:3–34), miracle stories (4:35–5:43), and anecdotes (8:27–38, 10:35–45, 12:13–34). The content of these stories tells the listeners of the amazing wisdom, authority, and power of Jesus, and makes plain to them his exemplary behavior on every occasion. It is all perfectly conventional. It remains the case, as occasionally with other Hellenistic "lives," that from time to time something else breaks through the convention. Through Mark's anecdotes we glimpse something that Mark cannot or does not tell us in so many words: we glimpse a real and utterly surprising person who drew men and women to him, and led them in his presence to a unique experience of the presence, power, and compassion of God.

Sources and Units of Composition

Soon after the opening of his *Demosthenes*, Plutarch talks about the conditions under which he would like to work:

> when one has undertaken to compose a history based upon readings which are not readily accessible or even found at home, but in foreign countries, for the most part, and scattered about among different owners, for him it is really necessary, first and above all things, that he should live in a city which is famous, friendly to the liberal arts, and populous, in order that he may have all sorts of books in plenty, and may by hearsay and enquiry come into possession of all those details which elude writers and are preserved with more conspicuous fidelity in people's memories. (*Demosthenes* 2.1, trans. Bernadotte Perrin)

Plutarch's intention here is actually to win our understanding for the fact that his knowledge of Latin is not (in his view) all that it should be, since he has chosen to live "in a small city . . . that it may not become smaller still" (2.2); but in the process of this narration he gives us a useful summary of the kinds of resource which he thinks important for a scholar.

Littérateur though he is, for some matters Plutarch still considers personal reminiscence and oral tradition the best source of all—and takes for granted that we shall agree with him. In so doing, of course, he simply follows the views of Herodotus and, more important, Thucydides, which were still accepted by virtually all who chose to write about the past (compare Arnaldo Momigliano, *Studies in Historiography*, 214). All writers of Hellenistic "lives" used oral sources. Cornelius Nepos refers in passing to oral tradition (*Cato* 1.1), and knew the subject of his *Atticus* personally (*Preface to Lives of Foreign Generals* 1). Tacitus was writing about his father-in-law, and Lucian about his former teacher, so naturally they drew upon reminiscence (*Agricola* 4; *Demonax* 1). Suetonius had been Hadrian's secretary and was friendly with Pliny; so it is not surprising that on more than one occasion he appeals directly to his own observations (*Augustus* 7.1; *Nero* 57.2; *Domitian* 12.2), or to hearsay (*Caligula* 9.3; *Otho* 10; *Titus* 3.2). As well as drawing upon Scripture, Philo makes extensive use of Jewish interpretative tradition, much of which he is likely to have acquired orally (for example, *On the Life of Moses* 1.13–14, 19, 21–33, 63–64).

Nonetheless, Plutarch's description of the writer's needs makes it clear that he also regards "all sorts of books in plenty [βιβλίων τε παντοδαπῶν ἀφθονίαν]" as important. His own writings reflect not only the wide and sensitive reading in Greek literature which had formed the basis of his education, but at times (as in the case of some of the Roman lives) also seem to provide evidence of research specifically undertaken for these projects (Christopher Pelling, "Plutarch's Method," 74 and passim). Tacitus seems to have drawn upon Agricola's own notes, and refers to ancient and modern writers (for example, *Agricola* 10). Suetonius used his subjects' own writings—Tiberius's autobiography, Claudius's memoirs, even Nero's poetry (*Tiberius* 61.1; *Claudius* 41.3; *Nero* 24.2). For subsequent scholarship, perhaps most important of all, he used the letters of Augustus (for example, 51.3, 64, 71); in addition, although perhaps with less zest, he also used historians (*Caligula* 16.1; *Tiberius* 61.3). On occasion—as in his discussion of Caligula's birthplace—we can see Suetonius at work, weighing and comparing his sources, particularly when they conflict (*Caligula* 8.1–5).

So much is true of the major biographers of our period. What of lesser writers? The *Life of Secundus* consists of four parts: (1) a novella (the fateful reunion of Secundus with his mother, which provides the reason for his silence), (2) a passion or martyrology (showing Secundus willing to die rather than abandon his principles, but not, in this case, leading to the hero's death), (3) a diatribe addressed by Secundus to Hadrian, and (4) a question-and-answer dialogue between Secundus and Hadrian (with the emperor, of course, providing the questions, and Secundus providing—in writing—the answers). Each of these parts involves a form with numerous literary parallels; thus some combination of written and/or oral sources seems likely to lie behind *Secundus*, although it is hard to say much more than that without an analysis more detailed than would be appropriate here.

The more sophisticated the writer, generally the less obvious are the junctures between the particular sources and the units of tradition that make up the work. Tacitus's and Plutarch's narratives usually flow as elegant and continuous wholes. In the *Life of Secundus*, by contrast, the four pieces stand simply side by side, and are easily identified. Yet even a sophisticated writer like Lucian may choose to structure a work very simply; *Demonax* is essentially a collection of anecdotes strung together in no particular order in a biographical framework.

In Mark's use of sources, I place him somewhere between Plutarch and Tacitus on the one hand, and *Demonax* and *Secundus* on the other. On the one hand, while there is by no means universal agreement, critics have long recognized the likelihood that previously assembled collections lie behind the Markan text. Among them are the five conflict stories (2:1–3:6); the parables (4:1–34); and the four mighty acts showing Jesus' mastery over the elements, the demons, disease, and death (4:35–5:43). Such collections appear to have been formed on the basis of common themes: others perhaps were based simply on catchwords (10:49–50). On other occasions what seem to have been originally discrete oral traditions have been used to construct literary forms. One example is the setting for the farewell discourse of Mark 13. Here we have an introductory peripatetic dialogue (13:1–2) followed by a seated conversation and discourse. This is a pattern to be found in a number of Greco-Roman dialogues (for example, Plutarch, *On the Obsolescence of Oracles* 5–6). A well-known (if slightly confused) Christian example is Justin's *Dialogue with Trypho* (1.1, 9.3, but compare 9.2).

On the other hand, while Mark does not combine his materials into a continuous whole with anything like the grace of a Plutarch or a Tacitus, still he does make considerably more effort in this direction than does the writer of *Secundus*. We have already referred to Mark's efforts toward creating a chronological framework. We should also note that his narrative provides us with summaries: general statements about what Jesus said and did that go beyond the accounts included. Some of these serve to introduce specific anecdotes (for example, 2:1–2, 4:1–2, 10:1); others conclude them (for example, 1:28, 4:33–34, 6:12–16); and others simply enable the narrative to run more smoothly (1:14–15; 32–34, 39; 3:7–12, 6:6b, 34, 53–56, 10:1, 15:40–41). Moreover Mark's narrative, including the "collections," is bound into a whole by the presentation and development of various themes, such as Jesus calling followers to himself (1:16–20, 2:13–14, 3:12–19), or Jesus enjoining silence about himself (1:25, 1:43, 3:12, 5:43). Mark's arrangement and development of themes will be of particular interest to us when we come to consider his use of oral techniques. For the present, we simply note that, as do many other writers of ancient "lives," he seeks to combine various sources into a narrative whole, and is by no means simply a collector of existing material.

Common Motifs

We have observed how certain genres are marked by the presence of particular motifs, such as the final shootout in the western. The ancients would have called these τοποι (transliterated, *topoi*). They are obviously a useful indicator of genre, although (as we have also observed) they need not all be present in any given example. Among recurring *topoi* in Hellenistic "lives" we should note the following.

(1) *The origins and birth of the hero* were invariably presented as indicators of what was to come. Suetonius, amid a catalogue of prodigies surrounding the birth of Augustus, tells of one that "gave warning that nature was pregnant with a king for the Roman people; thereupon the senate in consternation decreed that no male child born that year should be reared; but those whose wives were with child saw to it that the decree was not filled" (*Augustus* 94). Similarly, in describing the birth of Alexander, Plutarch tells us of mysterious dreams and other hints of divine begetting that accompanied it (*Alexander* 2–3).

Along with prodigies go the hero's lineage and background, and details of these remain important even when there are no records of prodigies. Thus, Philo has nothing miraculous to report about the birth of Moses—surprisingly, since such traditions were certainly current (compare Josephus, *Antiquities* 2.205)—but he is careful to speak of Moses' excellent lineage (*Moses* 1.5, 7); similarly Lucian with Demonax (*Demonax* 3), and Tacitus with Agricola (*Agricola* 4). Every one of Suetonius' *Lives of the Caesars* that is complete includes details of the subject's ancestry and descent. Sometimes, where Suetonius is particularly admiring of his hero, we notice him anxiously explaining away possible skeletons in the family cupboard (*Augustus* 1–4); with others whom he does not admire, we find him equally anxious to drag the skeletons out (*Tiberius* 2.1–4; *Nero* 2–5). Plutarch's "Lives" invariably include lineage (for example, *Alcibiades* 1; *Coriolanus* 1). An obvious exception to this pattern is, however, the anonymous *Life of Secundus*, which contains notice neither of its hero's birth, nor any detail of his lineage. This may be because the opening section of the "Life"—The Fateful Reunion—involves the humiliation and disgrace of Secundus's mother.

Formally, Mark is in this respect like *Secundus*; yet it must also be said that Mark's presentation of Jesus as the fulfillment of ancient prophecy, and the description (whether Mark's or another's) of Jesus as "Son of God" give some of the effects of the more usual details of lineage (1:1–1:8). Perhaps (in view of the somewhat deprecatory references to Jesus' family background in 3:21, 31–35 and 6:3) they represent the best that the evangelist felt able to do with the tradition he had received. It is striking, nonetheless, that in this respect Matthew and Luke both, in their different ways, bring Mark's narrative more into line with other Hellenistic "lives" (Matt. 1:1–25; Luke 1:26–27, 3:23–37).

(2) Following their accounts of the subject's birth and descent, the writers of "lives" will often proceed to reminiscences of *education or youthful exploits*. Lucian recalls how the philosopher Demonax,

> even from his boyhood felt the stirring of an individual impulse toward the higher life and an inborn love for philosophy, so that he despised all that men count good, and, committing himself unreservedly to liberty and free-speech, was steadfast in leading a straight, sane, and irreproachable life . . . (*Demonax* 3, trans. A. M. Harmon)

and then gives a general outline of the philosopher's education, without, however, any specific anecdotes (*Demonax* 4). In *Alexander*, Plutarch recounts specific youthful exploits such as Alexander's entertaining the Persian envoys, and his taming of his horse, Bucephalus (Plutarch, *Alexander* 1.2–8.4; compare *Cicero* 1.1–4.5 and *Demosthenes* 4.1–5.5). The *Life of Secundus* tells us briefly that Secundus was sent away for his education, that his father died, and that he learned contempt for women, before recounting, as the first act of his maturity, his return home and "fateful reunion" with his mother—a narrative both describing and explaining the silence that (aside from wisdom) is henceforth his most striking characteristic (*Secundus* [ed. Perry] 68.4–69.15).

Mark obviously lacks even such details of his subject's upbringing as the author of *Secundus* is able to provide. But Mark does provide us with two accounts of what we may broadly categorize as Jesus' early maturity: the stories of his baptism and testing (1:9–13). These stories perform functions in the narrative precisely akin to the youthful exploits that we have been considering in other "lives." They give early indications of the kind of person Jesus is, and prepare us for what is to come.

(3) *The deeds of the hero* form the basic stuff of any "life." Naturally, they are deeds appropriate to the kind of hero being described. Tacitus speaks of Agricola's ability as soldier and statesman. Suetonius, while often avoiding major political and military matters (which he probably regarded as more suitable for a "history") spends much time on the personal characters and tastes of the autocrats whom he describes, as well as on their running of the everyday business side of imperial administration. Lucian and the writer of *Secundus* fill the bulk of their narrative with material appropriate to the life of a philosopher: his conversations, responses to questions, and opinions. In just the same way, the deeds and words of Jesus that form the bulk of Mark's narrative illustrate the person of one who is to be regarded as sage, prophet, and Son of God.

(4) Finally we consider *the death of the hero*. We have already noted the extent to which the deaths of notable persons occupy attention in the literature of our period. As prodigies sometimes mark the births of the great, so they frequently accompany their deaths. Plutarch, describing Caesar's end, tells of several "events of divine ordering," including a comet that appeared for seven nights and an eclipse of the sun, all of which showed that Caesar's murder "was not pleasing to the gods" (*Caesar* 69.3, 5). According to Suetonius, Augustus's coming demise, "and his deification after death, were known in advance by unmistakable signs" (*Augustus* 97), and indeed, the deaths of Suetonius's emperors (good and bad) are almost invariably marked by prodigies and omens (for example, *Claudius* 46; *Tiberius* 74; *Caligula* 57; *Nero* 46; *Galba* 18). Moreover, Suetonius usually gives details of the death itself, including memorable last words.

The death of the philosopher Demonax is not, as one would expect of Lucian, marked by any prodigy, but it is graceful and appropriate: "he took leave of life with the same cheerful humour that people he met always saw him in" (*Demonax* 65). It is appropriate for sages, like Socrates, to suffer with dignity at the hands of uncomprehending or malicious society and its representatives, and while in the cases of Demonax and Secundus these events are not linked to their deaths, it is notable that both of them undergo such "passions"—the former being explicitly compared to that of Socrates (*Demonax* 11; *Secundus* [ed. Perry] 70.16–74.15). Subsequent events include the funeral, and any special honors afforded the hero (*Moses* 2.291; *Demonax* 67).

The extent to which Mark's account of Jesus' death matches those of other Hellenistic "lives" is striking. In Mark, too, the death is decribed, including a closing word (15:34–37). It is accompanied by prodigies in nature (15:33, 38). Jesus, like Socrates, has manifestly suffered with dignity at the hands of uncomprehending or malicious society and its representatives (14–15, *passim*). The burial of Jesus is described (15:42–47).

In two ways, however, Mark's account of Jesus' death departs strongly from other extant Hellenistic "lives" of the period. Mark departs from convention first, in the nature of the death itself, which is disgraceful; and second, in the closing scene of revelation, with its affirmation of the risen Jesus. Some parallel to the latter is provided by Philostratus in the appearances after death of Apollonius to his doubting followers, and his possible ascent into heaven (*Apollonius* 8.30–31), and we have already recalled the spoofed "resurrection" in *Peregrinus*. In both these cases, however, if we were to establish a connection with Mark, it would be with Mark as source, not recipient. More important, however, than our observation of such uncertain parallels is for us to note once more what we have already perceived in other connections. Points where the evangelist appears to depart from the genre are precisely those at which he was manifestly constrained by the tradition.

Written Style

Most Greco-Roman biographies that have survived are (as is not surprising) written with some literary pretensions, in language that is "periodic" (that is, with elaborate sentences); but it is clear that many others were written for popular consumption in an easier "paratactic" style (that is, with simple sentences set side by side). Surviving examples of the latter include various lives of Greek poets such as the *Life of Pindar* and the *Life of Secundus*. Also relevant here are those other surviving examples of contemporary narrative prose, obviously designed for popular consumption, that we mentioned earlier, such as Chariton's *Callirhoe*. A paratactic style is, of course, characteristic of the gospels, and it is highly likely that the general educational and social level for which they were written is also that for which *Secundus* and *Callirhoe* were intended. Students with enough knowledge of Greek to cope with the gospels, and not much more, may discover this for themselves by a simple experiment. If they attempt to read, say, Plutarch or Philo, they will probably find it hard going; if, by contrast, they undertake *Secundus*, Chariton, or Xenophon of Ephesus, then (save for occasional problems of vocabulary) they will most likely find the work easy enough (see Appendix).[2]

It was common at one time to speak of Mark's Greek as "Jewish" or Semitic, reflecting Hebrew or Aramaic idiom, either directly or through the influence of the Septuagint. Certain features of Mark's style were particularly identified as his "Semitisms," notably (1) asyndeton (the habit, unusual in Classical Greek, of omitting the particle from the beginning of sentences); (2) the common use of the historic present in narrative (for example, 1:12, 2:10, hidden in RSV but compare KJV); and (3) parataxis. Of parataxis we have already spoken; as for asyndeton and the historic present, it is evident that both were perfectly acceptable in various kinds of *koinē* Greek at this period, including many where there is not the slightest possibility of Semitic influence (compare Longinus, *On the Sublime* 18.1–2, 19).

It remains true that Mark's written style is among the least literary of the New Testament, and we must assert this, despite Mary Beavis's attempt in certain respects to defend him (*Mark's Audience*, 42–46). Consistent with this is the way in which Matthew and Luke constantly, in their different ways, seem to represent attempts to improve Mark's written Greek. So, incidentally, do many of the variants that characterize the textual tradition of Mark itself: Codex Washingtonianus, for example, appears repeatedly to omit phrases that the scribe felt were redundant or wordy (for example, 12:44, 14:30, 14:60; compare 14:61).

In contrast, while we must admit the generally nonliterary quality of Mark's Greek, we must not exaggerate the significance of that admission. It is perfectly clear (as we have already indicated) that there was at this period a wide range of speech available to the masses and to

2. It is, I think, extremely unlikely that we are here dealing with the deliberate affectation of popular idiom in order to create an effect of simplicity or vulgarity, as in J. D. Salinger's *Catcher in the Rye*. What we are dealing with in Mark and much of the rest of the New Testament is (despite Augustine's and Origen's suggestions to the contrary) a genuine popular literary style.

the well-educated alike. Mark's Greek is the language of popular written style (which tends to be close to spoken language) rather than that of the literati. It could have been read aloud to good effect, and would have been understood by everyone present, whatever their level of education.

Moreover, while Mark's Greek may be grammatically simple, his written style is by no means lacking skill in other respects. Rather, as Augustine Stock has justly observed, "The gospel abounds in picturesque details and lifelike suggestions: an expressive gesture or impressive look caught by Mark's pen, or a mood described by a relevant verb, or details of setting given in passing" (*Call to Discipleship*, 72). We recall the vivid picture of fishermen "casting a net in the sea" immediately echoed in the call "I will make you fishers for people!" (1:17); or the fierce vigor of demonic conflict in the synagogue at Capernaum: "Be silent, and come out of him!" (1:25). We think of Jesus, challenged over his authority to forgive sins, responding to his critics, and as he does so, turning to the paralytic to offer that healing which is the sign of forgiveness (2:10: despite the objections that have been raised by modern critics, the shift involved in "he said to the paralytic" [λέγει τῷ παραλυτικῷ] is perfectly permissible Greek—Luke and Matthew do not appear to have had any problem with it—and is a gift for anyone reading aloud). We think of Jesus, again in the synagogue, looking round at his critics, "grieved at their hardness of heart" (3:5). We remember him "asleep on a cushion" while the storm rages (4:38), marveling at unbelief (6:6), looking at the crowd with compassion (6:34), looking up to heaven and sighing (7:34), rebuking Peter (8:33), taking children into his arms (10:16 cf. 9:36), striding ahead to Jerusalem (10:32), looking round "at everything" in the Temple (11:11), sitting down opposite the treasury and watching (12:41), and, on that grievous last night, taking the bread, and blessing, and breaking, and giving, as he has done before (14:22; compare 6:41, 8:6).

Such effects as these suggest the art of one whose concern is not polished prose, but effective narrative—and, what is more, effective narrative *performed*. Such effects suggest one who, as Stock observes, "in the course of his discourse, can stress a point with a motion, a silence, or an expressive look" (*Call to Discipleship*, 73). Indeed, that Mark's gospel was primarily created for reading aloud would partly explain its tendency to redundancy; even the stylist Demetrius admits, "For the sake of clearness the same word must often be used twice.

Excessive terseness may give greater pleasure, but it fails in clearness. For as those who race past us are sometimes indistinctly seen, so also the meaning of a sentence may, owing to its hurried movement, be caught only imperfectly" (*On Style* 4.197, trans. W. Rhys Roberts). In any case, it is evident that Mark's Greek is no worse than that of *Secundus*, and comes well within the range of literary styles that was possible for a Hellenistic "life."

Length

Richard Burridge rightly insists that when discussing genre we pay attention to the question of length, noting how often "this feature has been ignored by scholars, with the consequence that works of quite differing sizes, and therefore possibly different genres, are lumped together with the gospels" (*What Are the Gospels?* 117–18). Aristotle certainly regarded size (*megethos*) as one characteristic of the tragic genre (*Poetics* 1449a 19, 1450b 26), and Alastair Fowler goes so far as to claim that "a genre not characterized by any definite size is not a kind" (*Kinds of Literature*, 64; cited Burridge, ibid.)

Hellenistic supper parties and other social occasions often included reading as an entertainment, and "lives" will have had their place among the pieces chosen (Pliny, *Letters* 1.15). Most of the "lives" we have been considering are of a length that would have been suitable for this, ranging from the *Life of Secundus*, occupying about a fifth of a scroll, to Plutarch's *Alexander*, requiring an entire scroll. The former could have been conveniently read to a group, perhaps with a couple of other pieces of similar length, at a single sitting. The latter would perhaps have been read at a single sitting, or could even have been the substance of two sittings. Into this range, Mark fits well. It could have been read to a group at one sitting in slightly under two hours—not an excessive length of time by the standards of the period, as we shall see. Breaks might be made at 8:21 (where there is a particularly powerful rhetorical climax) and at 10:45 (the eve of the final "act" in Jerusalem), allowing for a performance in three "acts," the first of about fifty minutes, the second of about twenty, and the last of about forty-five, with short intervals between.

Hellenistic supper parties frequently, if not normally, had religious connotations. In antiquity, as Ramsey MacMullen has pointed out,

> For most people, to have a good time with their friends involved some contact with a god who served as guest of honor, as master of ceremonies, or as host in the porticoes or flowering, shaded grounds of his own dwelling. For most people, meat was a thing never eaten and wine to surfeit never drunk save as some religious setting permitted. There existed—it is no great exaggeration to say of all but the fairly rich—no formal social life in the world of the [second-century] Apologists that was entirely secular (*Paganism in The Roman Empire*, 40).

So we may easily imagine Mark's "Life of Jesus" being read in connection with the Christian Eucharist. It would naturally have found its place in Christian households, which were the nuclei of local churches (Rom. 16:5; 1 Cor. 16:19; Philem. 2), centers for assembly and perhaps also bases for faction (1 Cor. 1:11–16; 3 John 9–10). It could have functioned in such settings as the Haggadah for a Christian Passover, as Augustine Stock and others have suggested (*Method and Message of Mark*, 12–19). However, it might equally well have been read entire in any Christian assembly. We have only to reflect on accounts of the Eucharist such as that in Acts 20 to see that such gatherings were obviously not, in general, expected to be hurried affairs. On this occasion, it will be recalled, Paul talked so long that poor Eutychus dropped off to sleep, fell out of a window and was lucky not to break his neck; even then, after the fuss, the group went on until daybreak (Acts 20:7–12). Clearly not much has changed (at least in this respect) seventy or so years later when Justin, in his description of the Christian Eucharist, tells how, "on Sundays, there is an assembly of all who live in towns or in the country, and the memoirs of the apostles or the writings of the prophets are read for as long as time allows" (*Apology* 1.66–67). We have referred to the Christian (and general) practice of public teaching, lecture, and discussion: for so long as Christianity hoped to be regarded as a "permitted religion" (*religio licita*) public readings of a "life" of the founder could have found their place in that context, also.

Function

The reader may recall the passage from C. S. Lewis quoted near the beginning of this book. Having spoken of our need, in judging any artifact, to know "*what* it is," Lewis at once qualified that as meaning "what it was intended to do and how it is meant to be used." It remains, then, for us to consider Mark's function.

> Aut prodesse volunt aut delectare poetae
> aut simul et iucunda et idonea dicere vitae
>
> [Poets desire either to benefit or to amuse,
> Or to speak words at once pleasing and helpful for life]
> (Horace, *Art of Poetry* 333–34).

This is the classical ideal: that art shall both entertain and edify, offering a blend of delight and instruction. Hellenistic "lives" were certainly designed with these ends in mind. The setting in which they were read was, as already observed, frequently a social gathering such as a supper party, where, after the benefit of a meal and in company with friends and the god, the stimuli of entertainment and education were appropriate. Naturally, some writers tend more to the one than to the other. Plutarch is the more concerned with edification; Lucian wants mostly to entertain. Yet it is obvious that even Lucian would claim (like Jack Point) to give a grain or two of wheat among the chaff, and it is equally clear that the learned and edifying Plutarch also intends to interest and attract—in a word, to entertain.

The Written Gospel as Entertainment

We are not used to reflecting on the written gospel as entertainment; yet we shall rob Mark of much of the credit due him, indeed, we shall deny him his seriousness of purpose, if we overlook that in choosing to present the "matter" of Jesus as a "life," he chose a form in which a great deal of material about his Lord (words and works, passion and resurrection) could be presented effectively and attractively. And we shall probably do less than justice to his intelligence if we suggest that *he* overlooked it. It is much more likely that he wrote as he did because he hoped to keep even Eutychus awake. In this connection we should note in particular the text's frequent asides. All of them are explanatory: they explain Aramaic words (6:41, 7:11, 34, 15:22, 34), Jewish customs (not always entirely accurately) (7:3, 14:12),[3] and even the

3. Mark's attempts to explain things Jewish have been seen by some as evidence that he must have written for a predominantly gentile audience. The conclusion displays, I think, a certain lack of rhetorical awareness. Effective speakers or storytellers will try to carry

story itself (2:10b, 7:26, 12:42). All of them offer a reader natural opportunities to keep *all* the members of the audience *au fait* with the action, and hence (it is to be hoped) awake. If we, however, find Mark's work less entertaining than was intended, this may be a further indication that we have ceased to use the gospel as it was designed to be used, as a spoken whole. When, by contrast, some critics attacked Alec McCowen's theatrical performance of Mark as not theologically responsible, one was left with the curious feeling that the real problem may have been that in performance some of Mark actually turned out to be rather funny—that the "street smarts" and repartee by which Jesus outwits and discomforts his opponents sometimes lead us, in our appreciative enjoyment, to smile. Of course, there are always questions of appropriateness and taste to be considered. Yet in general terms it must be said that we cannot accept an absolute opposition between humor and high seriousness: witness Congreve, or Alan Ayckbourn. Or even *Macbeth.*

The Written Gospel as Edification

Hellenistic biographers, like Hellenistic historians, tended to regard the past as a source of useful lessons for the present. Above all, to write a "life" was to claim that the career of an historical personage was worth remembering: there are no "lives" of legends. Hellenistic "lives" are often encomiastic, presenting their heroes as models of virtue: particularly, of course, the virtues appropriate to their type, and therefore as suitable for emulation. Even when, as in *Demetrius* and *Antony*, Plutarch narrates the doings of evil men, he is insistent that he does so not merely to "divert and amuse my readers by giving variety to my writing" (*Demetrius* 1.5); rather, he will illustrate morality by its opposite (and,

---

along their *whole* audience—and that may well mean inserting explanatory asides if even a few of those present will be helped by them. Provided this is done briefly (and Mark's asides are always brief) it will not adversely affect the attention of the better-informed: indeed, being made thus aware that one is better informed than others is not necessarily unpleasant. (Of course, it remains that Mark *may* have written for a predominantly gentile audience: I merely observe that such a conclusion needs to be established on other grounds.) By contrast with his attempts to explain things Jewish, Mark apparently sees no need to explain Latin loan words (see 5:9, 6:37, 12:15, 15:39)—as has often been noted. Yet this, too, hardly tells us much about his audience: such words as *legio*, *denarius*, and *centurio* will have been generally known throughout the Empire.

incidentally, show that neither vice nor virtue is the property of one nation). No doubt this insistence on the ethical value of history was partly because the ancients lacked what we should call a historical critical sense: they assumed an identity between themselves and the past that we should now question. Be that as it may, the fact that the past provided lessons for the present was, in their eyes, one important reason for studying it. That Mark's Jesus is seen as a paradigm for emulation generally goes without saying: yet occasionally that, too, is spelt out, notably at Caesarea Philippi (8:34–38) and on the road to Jerusalem when the disciples are quarreling over their places in the Kingdom (10:42–45).

More specifically, ancient biography was used at times as a tool for propaganda: to support or discredit a view, a movement, or an individual. The creation of biography could be a serious political act. In first-century rhetorical terms, "lives" were mainly epideictic (that is, seeking to persuade their auditors to hold or reaffirm a point of view in the present, and celebrating a particular person), but they might have deliberative elements (that is, they might seek to persuade them to action in the future). Cato the Younger was the subject of a propaganda war, with admiring pamphlets about him by Cicero (106–43 B.C.E.), Marcus Brutus (85–42 B.C.E.), and others, and hostile ones (no longer extant) by Julius Caesar (100–44 B.C.E.) and Augustus (63 B.C.E.–14 C.E.). Under the tyrant Domitian (81–96 C.E.), to have written a biography of Helvidius Priscus (who had been critical of the principate) was a capital offense, and cost Herennius Senecio his life (Tacitus, *Agricola* 2).

Mark's intention to speak of "Jesus Christ, Son of God" (1:1) clearly falls within these categories. He too, in rhetorical terms, is mainly epideictic, though with deliberative elements (8:35–38, 10:42–45, 13:5–37). The gospel is not, of course, a "missionary pamphlet" in our sense. The mere circumstances of first-century book production precludes such a notion. It would, as we have observed, have been read, usually aloud, in Christian households and assemblies. But such households and assemblies no doubt included many whose commitment to the faith was marginal or doubtful. We should also bear in mind the possibility referred to earlier that the gospel was designed for more public readings. In any case, it is clear that Mark wishes his hearers to believe or to go on believing that Jesus is the anointed, the Son of God (1:1, 34, 8:29, 9:41, 12:35, 14:61–62, 15:32; 1:11, 3:11, 5:7, 9:7, 12:6, 14:61–

62, 15:39), and to lead lives that accord with this (1:17, 20, 8:34–38, 10:21, 10:29–31, 13:32–37).

In connection with this purpose, it is interesting to note the number of minor climaxes in Mark's narrative where Jesus turns from addressing his disciples to address a wider listening group with whom an assembled Christian household or similar assembly could naturally associate themselves—especially if assisted to do so by a sensitive and experienced reader. Such passages touch on precisely the kinds of issue with which a pastor or preacher would be concerned: they remind the hearers that if they try to be open to God's will, then regardless of their status in the world's eyes, they are Jesus' true family (3:34–35); that it is not religious externals that matter, but what comes from their hearts (7:14–23); that they must not be ashamed of Jesus's cross, but must take up their own cross and walk with him (8:34–38—note how the storyteller pictures Jesus summoning "the multitude with his disciples," making it easy for the reader to include the audience with the simplest of gestures and inflections, and then clearly has Jesus turn *back* to the disciples in 9:1—"and he said to *them*" as the story continues); that the community must be gentle with its weakest members, and members must be at peace with one another (9:42–50); that they must not attempt to dominate one another, but serve each other, as Jesus came not to be served but to serve, offering his life for them (10:42–45); that they must expect to endure sufferings, for the sake of the gospel (13:9–13; compare 8:34–38); and that they must stay awake, looking for the coming of the Son of man (13:37). (What a gift for a performer that last line would be: "And what I say to you I say to all: Watch!")

Preliminary Conclusion

In eight of eleven major features we have found Mark conforming exactly to what would have been expected of a Hellenistic "life." In two of the three features where he differs, we have found that he was manifestly constrained to do so by the tradition he had received. The tradition of Jesus as Son of God obliged Mark to stretch previous boundaries regarding the hero of a Hellenistic "life"; the tradition of Jesus' death by crucifixion, and his subsequent resurrection, obliged Mark to transgress the genre's previous boundaries regarding the death of the hero. We may now reasonably claim that, granted its unusual features, it is as an example of a "life" that Mark's text would have been received by any averagely educated Greco-Roman audience.

In the course of this discussion we have noted features indicating that the gospel was designed to be read aloud as a continuous whole in public performance. Of course this idea is by no means novel. Various scholars from various viewpoints have argued that Mark was meant to be used in this way (Boomershine and Bartholomew, "Narrative Technique of Mark," 221; Rhoads and Michie, *Mark as Story*, passim; Hengel, *Studies in the Gospel of Mark*, 52; Standaert, *L'Evangile selon Marc*, 3, 9, 26; Tolbert, *Sowing the Gospel,* 72, 82; Beavis, *Mark's Audience*, 124). Augustine Stock speaks of effects "proper to oral delivery [that] were apparently integrated into the gospel so that they could be renewed with each public reading" (*Call to Discipleship*, 73). These suggestions lead us directly to the second part of our study. Indeed, they begin already to suggest the answer to our second question.

BIBLIOGRAPHY

On the term "gospel," see Helmut Koester, *Ancient Christian Gospels: Their History and Development* (London: SCM; Philadelphia: Fortress, 1990), 1–48.

On Mark 1:1, see Peter M. Head, "A Text-Critical Study of Mark 1.1: 'The Beginning of the Gospel of Jesus Christ'," *NTS* 37.4 (1991): 621–29; for a different view, see Alexander Globe, "The Caesarean Omission of the Phrase 'Son of God' in Mark 1.1," *HTR* 75.2 (1982): 209–18.

On religious rhetoric, and rhetoric in Mark, see George A. Kennedy, *New Testament Interpretation Through Rhetorical Criticism* (Chapel Hill: University of North Carolina Press, 1984), especially 6–7, 104–7. Mary Ann Tolbert, *Sowing the Word: Mark's World in Literary Historical Perspective* (Minneapolis, Minn.: Fortress, 1989) has a helpful discussion of the rhetorical aspects of Mark's opening (108–13).

On *chreiai* and the gospels, see, for a brief treatment, Vernon K. Robbins, "The Chreia" in David E. Aune, ed., *Greco-Roman Literature and the New Testament: Selected Forms and Genres* (Atlanta, Ga.: Scholars, 1988), 1–23; a fuller treatment is offered by Vernon K. Robbins and Burton L. Mack, *Rhetoric in the Gospels: Argumentation in Narrative Elaboration* (Philadelphia: Fortress, 1987).

On issues of characterization in Hellenistic "Lives," see Christopher Gill, "The Character-Personality Distinction," in Christopher Pelling, ed., *Characterization and Individuality in Greek Literature* (Oxford: Clarendon, 1990), 1–31; also Pelling's own article "Childhood and Personality in Greek Biography," in the same book, 213–44. There are some useful general observations in Patricia Cox, *Biography in Late Antiquity: A Quest for the Holy Man* (Berkeley and Los Angeles: University of California Press, 1983), ix–xvi, 3–16. Despite the work that has been done on characterization in Hellenistic "lives," much New Testament scholarship continues to follow Bultmann in asserting that the gospels cannot be "biographical" because they show no interest in Jesus's "human personality" (*The History of the Synoptic Tradition* [New York: Harper and Row, 1963], 372); among recent examples may be cited E. Eugene Boring, "How Gospels Begin," in Dennis E. Smith, ed., *How Gospels Begin*, Semeia 52 (Atlanta, Ga.: Scholars, 1991), 64, as well as the assumptions behind the entire debate between Charles H. Talbert and David Moessner on the relationship of the gospels to "the essence of the person" (Talbert, "Once Again: Gospel Genre," Semeia 43 [: Scholars, 1988], 53–73; Moessner, "And Once Again, 'What Sort of Essence?' A Response to Charles Talbert," Semeia 43 [Atlanta: Scholars, 1988], 75–84.)

On sources and units of composition, see Christopher B. R. Pelling, "Plutarch's Method of Work in the Roman Lives," *JHS* 99 (1979): 74–96; idem, "Plutarch's Adaptation of his Source Material," *JHS* 100 (1980): 127–40; A. Wallace-Hadrill, *Suetonius: The Scholar and His Caesars* (London: Duckworth, 1983), especially 50–96. For discussions of Mark's sources, see Helmut Koester, *Ancient Christian Gospels: Their History and Development* (London: SCM; Philadelphia: Trinity, 1990), 286–89; for a somewhat different view, see Augustine Stock, *Call to Discipleship: A Literary Study of St. Mark's Gospel* (Wilmington, Del.: Glazier, 1982), 45–46.

On connections between the language of the masses and the language of the formally educated, see the bibliography for Chapter 2. On Mark's written style, see Vincent Taylor, *The Gospel According to St. Mark* (London: Macmillan, 1957), 44–54. Herman C. Waetjan, *A Reordering of Power: A Socio-Political Reading of Mark's Gospel* (Philadelphia: Fortress, 1989) attempts to convey in English the effect of Mark's popular style by means of a translation of the complete gospel (27–61; compare 2–3, 17–18). On the Markan textual tradition represented by Codex Washingtonianus as "improving" Mark's style, see Larry W. Hurtado, *Text-Critical Methodology and the Pre-Caesarean Text: Codex W in the Gospel of Mark* (Grand Rapids, Mich.: Eerdmans, 1981).

On the gospel's being designed to be heard as a single narrative, see David Rhoads and Donald Michie, *Mark as Story: An Introduction to the Narrative of a Gospel* (Philadelphia: Fortress, 1982), which also provides an interesting introduction and explanation to Rhoads's videotape (see "An Audio-Visual Resource" below. Also of interest (though in my view wide of the mark) is John Koenig, "St. Mark on the Stage: Laughing all the Way to the Cross," *Theology Today* 36 (1979): 84–88, which is highly critical of a stage performance of Mark by British actor Alec McCowen; see also *Time*, March 10, 1980, 65, for some of McCowen's own observations.

On the gospel as entertainment, see Mary Ann Tolbert, *Sowing the Word: Mark's World in Literary-Historical Perspective* (Minneapolis, Minn.: Fortress, 1989), 72. Compare also Tomas Hägg, *The Novel in Antiquity* (Oxford: Basil Blackwell, 1983), 90–96; the section suggests a number of interesting parallels with the probable Markan milieu.

An Audio-Visual Resources

Alec McCowan's performance of St. Mark's Gospel in the King James Version is available on videotape; it is distributed by the American Bible Society, 1865 Broadway, New York, NY 10023. Compare David Rhoads, *Dramatic Performance of Mark*, distributed by Select, 2199 E. Main Street, Columbus, Ohio. Rhoads is not as good an actor as McCowan; nonetheless he has occasional flashes of insight, and his translation (prepared for the purpose) is at times interesting.

II

WAS MARK WRITTEN TO BE READ ALOUD?

6

Orality and Oral Transmission

Did Mark write for a *listening* audience? In what ways might that have affected his work? Recent studies of orality and its relationship to the rise of literacy have provided us with useful tools with which to examine such a question, and it is in the context of these studies that I now wish to consider Mark. Like all who enter this field, I shall be drawing continually on the foundational researches and insights of the great Homer scholar Milman Parry and his student Albert Bates Lord, together with the further work of Eric A. Havelock and Walter Ong, S.J.

Ong, in a number of books and articles (for example, *Presence of the Word*, 1–110), has conveniently distinguished four stages of psycho-cultural development in human societies:

1. primary oral culture, which is largely or completely innocent of reading and writing
2. literate culture in the manuscript stage, for example, from ancient Greece through medieval Europe; this culture is literate in that it uses phonetic alphabetic writing
3. literate culture in the print stage
4. secondary oral culture, springing from use of electronic media, but rooted in literacy

It is, of course, stages 1 and 2 in this sequence that are of particular interest to students of the Bible. The Hebrew Scriptures seem to provide examples of exactly the kind of "oral traditional literature"—that is, textualization of originally preliterate oral tradition—whose study was undertaken by Parry and Lord. With the New Testament, however, the position is more complicated. Obviously, the milieu that produced the New Testament was well into the second of Father Ong's stages: that is to say, it was literate; indeed, highly so. Perhaps in no other society before those of late nineteenth-century North America and Europe was literacy so widespread. Nevertheless, the tempting conclusion that the study of oral modes of thought and composition is irrelevant in studying Mark would be mistaken, for at least three reasons.

First, with regard to Ong's stages, we should remember as a general truth that there neither was nor is in practice any clear-cut or absolute division between them. People (individuals or communities) do not become literate and cease to speak or listen. People do not begin to use print and cease to read manuscripts or write by hand. Even the users of electronic media continue to use other media. Thus, even in the most technologically advanced societies, people still learn physical tasks (like making bread or holding a child), sporting activities (like jumping or swimming), and artistic pursuits (like singing or playing an instrument), exactly as people have always learned such activities—by familiarization, by spoken word, and by example. Therefore, in describing a society's shift from, say, oral to literate culture, we should never think of complete change, but rather, as does Lou H. Silberman, of "a shifting of focus; the center, voice, became the fringe; the fringe, text, becomes the center; but both remain within the global experience" ("Reflections on Orality," 3).

Second, the change from oral to literate habits of mind, even so far as it does take place (and it certainly does), is not sudden, nor even swift. Homer remained the ancients' favorite poet. The initial effect on us of new technologies is invariably not so much to change what we do as to enable us to do more efficiently what we are doing already. (I continue to do with my computer exactly the things I used to do with my typewriter, only more easily—to the abiding scorn of my more "computer-literate" friends!) Only with usage—and "with usage," particularly in antiquity, can mean a very long time—do what we call "the possibilities of the medium" begin to show themselves. As regards literacy, this is particularly true of those who are only marginally or averagely literate: they, virtually by definition, are those who remain nearer to preliterate habits of mind and thought than those who are highly literate.

Third, and most important for our immediate study, even where second-stage literacy is highly developed, as with a Plutarch or a Lucian, it still remains very different from the third-stage literacy with which we are familiar; ours is a *print* literacy, and theirs was a *manuscript* or *chirographic* literacy. Chirographic literacy is bound to be closer to orality than print literacy; dependence upon manuscripts means that even if one's experience of poetry, story, or lore is no longer simply *oral*, it is still, inevitably, mostly *aural*.

For Jew and Gentile alike in the first Christian century, to read anything regarded as "literature" was normally to read aloud. Solitary reading was a privileged exercise, and "publication" was much more likely to mean public performance than the dissemination of copies. Such was the situation that Pliny, for example, took for granted; such was the situation reflected vividly in a letter to his friend, Cornelius Minicianus:

> Today is one day when I must be free. Titinius Capito is giving a reading, which it is my duty—or perhaps my urgent desire—to attend. He is a splendid personality who should be numbered among the shining lights of our generation; a patron of learning and admirer of the learned, whom he supports and helps in their careers. . . . He lends his house for public readings, and is wonderfully generous about attending those which are held elsewhere; at any rate, he has never missed one of mine, provided that he was in Rome at the time. (Pliny, *Letters* 8.12, trans. Betty Radice)

Most of those who experienced the poets or the Scriptures would experience them in performance: in theatre, hall, or synagogue, or perhaps in the family. To read aloud effectively from the unpunctuated texts of the period was a skill, moreover, requiring preparation and considerable familiarity; as Henri Marrou has shown us, the first aim of ancient education was to teach that skill (*History of Education in Antiquity*, 165). Among those who listened, many would no doubt themselves be illiterate. Therefore there was (as we have observed) still a degree of commonality in interests, ideas, and even speech between the highly educated and the illiterate that is quite hard for us to contemplate; and therefore most writers wrote to be *heard*.

So Hellenistic society in the first century of the Christian era, although no longer primarily oral, was still residually so. Authors of the period had begun to be dominated by the (comparative) fixedness of written text, and to that extent had moved away from oral habits of mind. But since for most people most of the time "texts" were still things heard rather than things seen, and "words" were sounds, not signs on a page, to that extent they had not moved very far. Hence their continuing preference for the reliability and precision of the spoken over the written word (for example, Plutarch, *Demosthenes* 2.1; Eusebius, *Ecclesiastical History,* 3.39.4) and their abiding passion for rhetoric—the art of public speaking—as the sine qua non of civilized education (for example, Philostratus, *Life of Apollonius of Tyana* 6.36). The noble Germanicus, possessing "all the highest qualities of body and mind," was to be admired not least because he "pleaded causes even after receiving the triumphal regalia [*oravit causas etiam triumphalis*]" (Suetonius, *Caligula* 3.2).

In the study of oral poetry, Paul Zumthor distinguishes between oral *tradition* (where such activities as composition, storage and repetition are all carried on orally) and oral *transmission* (which need only involve oral transmission and reception) (*Oral Poetry*, 22–23). He regards as essentially oral any poetic composition "where transmission and reception at least are carried by voice and hearing" (23). "Variation in the other operations" merely "modulates this fundamental orality" (23). Whether we choose to state the matter this way depends, of course, on how we decide to define the word "orality." In any case, we rest secure that what students of orality have to teach us about oral habits of mind and composition will be important in studying the literature of the Hellenistic period.

BIBLIOGRAPHY

For the study of orality, the following are foundational, and are best read in the order in which they are here set out: Adam Parry, ed., *The Making of Homeric Verse: The Collected Papers of Milman Parry* (New York: Oxford University Press, 1987); Albert Bates Lord, *The Singer of Tales* (Cambridge: Harvard University Press, 1960); Eric A. Havelock, *Preface to Plato* (Cambridge: Harvard University Press, Belknap, 1963); Walter J. Ong, S.J., *The Presence of the Word: Some Prolegomena for Cultural and Religious History* (Minneapolis: University of Minnesota Press, 1967). Paul Zumthor, *Oral Poetry: An Introduction* (Minneapolis: University of Minnesota Press, 1990) should probably be added to this list.

On the general subject of this chapter, see Ong, *Presence of the Word*, 1–110; Zumthor, *Oral Poetry*, 3–31; also Paul J. Achtemeier, "*Omne Verbum Sonat:* The New Testament and the Oral Environment of Late Antiquity," *JBL* 109.1 (1990): 3–27 (though on the question of reading aloud, see the qualifications offered by Bernard M. W. Knox, cited below).

On reading aloud in antiquity, see G. L. Hendrickson, "Ancient Reading," *CJ* 25 (1929): 182–96; but for some interesting qualifications, compare Bernard M. W. Knox, "Silent Reading in Antiquity," *GRBS* 9.4 (1968): 421–35. On learning to read, see Henri I. Marrou, *A History of Education in Antiquity* (London: Sheed and Ward, 1956), 151–54; also Stephen Bonner, *Education in Ancient Rome: From the Elder Cato to the Younger Pliny* (Berkeley and Los Angeles: University of California Press, 1977), 165–80.

On the importance of rhetoric in classical antiquity, see Marrou, 79–91, 194–205; also Bonner, 65–89, 250–327. See further George A. Kennedy, *Classical Rhetoric and its Christian and Secular Tradition from Ancient to Modern Times* (Chapel Hill: University of North Carolina, 1980).

7

Some Characteristics of Oral Composition

What were (and are) the characteristics of oral style? What marks the compositions by which orality transmits what is important? In this part of our study we shall not be bound by the criteria of time, place, and social level that limited us when we considered Mark's genre. Genres exist and are significant in particular milieus, and milieus, by their nature, imply temporal, spatial, and social boundaries. Our present concern is with the characteristics of *oral discourse*. Therefore any composition is relevant to that concern if it involves oral discourse: that is to say (accepting Zumthor's description) if "transmission and reception at least are carried by voice and hearing" (*Oral Poetry*, 23).

Orality and Narrative

Anything that is to be orally transmitted must be memorable. Preliterate orality, in particular, can ask to be told again, but it cannot look things up. Naturally a preliterate society determined that Mnemosyne (Memory) was mother of all the Muses. We remember that with which we can identify, and we identify most easily with persons, so oral knowledge must be situated in stories about people. This means narrative.

Thus, orality does not discuss the nature and benefits of faithfulness in marriage; it tells us

> How excellent was the mind of blameless Penelope,
> Daughter of Ikarios! How well she kept in mind Odysseus
> Her lawful husband! Thus shall never perish the renown
> Of his merit. The immortals shall fashion for mortals
> A pleasant song for sensible Penelope.
>
> (*Odyssey* 24.194–98)

Orality does not give us a treatise on the relative value of physical and mental ability, it tells us stories about Achilles, who had "might" (βίη) and Odysseus, who had "artifice" (μῆτις). Orality does not analyze the tensions between ritual and behavior, it tells us the story of God's relationship with Israel: "For I desire steadfast love and not sacrifice, the knowledge of God rather than burnt offerings" (Hos. 6:6). In the *Iliad*, even a notion such as "heavy fighting" means a series of vividly told stories of individual combats in which detailed (and endlessly varied) anatomical descriptions of fatal wounds are combined with pathetic little biographies of the dying losers. Occasionally the poet pulls back, and evokes the general scene, but still the verse is strongly visual, elaborate similes of rivers in flood or forest fires serving to bring before our mind's eye the ranks of warriors bearing down on each other across the plain.

Orality and Hyperbole

Actually, despite what Hosea said in the passage quoted above, God *did* require sacrifice (compare Lev. 1–7), and, as his second line suggests, Hosea knew it. Orality's precepts tend to hyperbole and exaggeration, rather than nuance or balance. "By keeping knowledge embedded in the human lifeworld, orality situates knowledge within a context of struggle" (Ong, *Orality and Literacy*, 44). The Book of Jeremiah does not provide us with a discussion of Israel's future and God's will; it tells us how the prophets Jeremiah and Hananiah hurled assertions at each other like weapons, and the loser died (Jer. 28; compare 1 Kings 22). Orality thrives on polarities and antagonisms, so it is not easy to argue with its claims. Auerbach's dictum about Holy Scripture tends to apply to oral style generally: "if we refuse to be subjected, we are rebels" (*Mimesis*, 15). This was Plato's main objection to orality (which for him was represented by poetry) as a way to knowledge. Plato wanted to *discuss* things (Havelock, *Preface to Plato*, *passim*).

Orality and Parataxis

Because oral transmission must be memorable, its narratives tend to have simple plots, and their structures are paratactic (setting events side by side) rather than complex (subordinating one event to another as cause and effect). "Action succeeds action in a kind of endless chain" (Havelock, *Preface to Plato*, 180). Naturally, if there is to be narrative at all, chronology and cause-and-effect will occasionally influence arrangement, simply on the basis that one thing happens after (and sometimes as a result of) another. As Vladimir Propp observed, "Theft cannot take place before the door is forced" (*Morphology of the Folktale*, 22). There is a basic conceptual unity between, say, an episode describing how a character decided to act in a certain way, and an episode that shows her subsequently doing so. Yet chronology and cause-and-effect are certainly not the only principles of arrangement in lengthy oral narratives. Nor could they be. In his *Art of Poetry*, Horace observed how the epic poet "hastens into the action and precipitates the hearer into the middle of things [*in medias res*]" (148–49). Horace had perceived one of the features of orally composed epic. It happens that this is a vivid and exciting way to begin, and catches our attention. Jane Austen was as well aware of that as Homer, and did it as effectively, in her own way. What Horace perhaps did not see was that for an oral poet to plunge *in medias res* was in any case the natural way of proceeding (Ong, *Orality and Literacy*, 144). No doubt Homer knew many songs and many varieties of episode about the Trojan War; but beyond the most obvious matters connected with its beginning and its end, there could be no way in which those episodes could have a chronology for him, nor, in the absence of writing, was there any way that a chronological list could exist. If something in an episode needed to be explained in the light of what had happened earlier, that would be dealt with in a flashback. So in the beginning of the *Iliad*, having begun to speak of the quarrel between Achilles and Agamemnon, Homer pauses, and turns aside to tell us what led up to it (*Iliad* I.8–100); that is the oral way of proceeding.

Orality and Formula

However, while many episodes in a long, orally composed narrative will be paratactic rather than linked by chronology or cause-and-effect, that is not to say that there is no principle in their arrangement. The guiding principle is recall. Lengthy narrative is managed in oral cultures largely by rhythm and formula. "The basic principle," observes Eric Havelock, "can be stated abstractly as variation within the same. . . . it preserves an identity, yet it makes room for a difference within this identity" (*Preface to Plato*, 147).

What Milman Parry and Albert Lord examined with great success was the formulaic vocabulary and expressions of first the Greek, and later of other oral epic poets—a formula being, in Parry's words, "a group of words which is regularly employed under the same metrical conditions to express a given essential idea" (Parry, *Making of Homeric Verse*, 272). Thus, Odysseus is "crafty Odysseus" (πολύμητις 'Οδυσσεύς) eighty-one times in Homer, not just because the adjective is appropriate and significant (though it is), but because with that epithet, the name Odysseus could readily be fitted into the meter, particularly when the character had to say something.

There is also rhythm and formula in the presentation of characters in oral narrative. They tend to be larger than life. The story of a hero frequently involves standard themes. He begins his career with testing, and the choice of companions. Often he is hidden or obscure, and then revealed. His adventures involve the assembly, the meal, the helper, the duel, and the marriage. At the end he is failed or betrayed by a close companion, deserted by others, and finally killed, yet not without glory or some hint of victory. Samson, Heracles, Theseus, Achilles, Beowulf, Roland, and Arthur all show elements of the pattern. (This is not to say that oral compositions follow a hero's entire career from beginning to end, in the manner of, say, Sir Thomas Malory's *Le Morte Darthur* as printed by Caxton. In general, they don't. They usually present a single episode, like that of Achilles before Troy. Malory himself had actually created a series of quite separate romances, and *not* a single book as Caxton implied [Eugène Vinaver, *Works of Sir Thomas Malory*, 1:vi]. Yet in oral narratives it is also perfectly clear that narrator and hearers are generally well aware of the whole "matter" of which they are considering a part.) Lesser characters, on the other hand, are often present in oral compositions only to play a part, and then to disappear. They have a role, but hardly personality. Think of those so-easily-interchangeable warriors who appear and disappear in the *Iliad*, like insignificant Pylaemenes, so forgettable that Homer even forgets that he has killed him! (*Iliad* 5.576–79; compare 13.643). In orality, as Propp pointed out, what matters is not delineation of the individual, but function (*Morphology of the Folktale*, 20–21).

Formula and rhythm are visible in narrative structure, too. Episodes may be marked by similarities (often called "correspondence") of content, form, or vocabulary—sometimes by all three. Such episodes are often arranged in patterns. One such pattern is a repetition of sequence, sometimes called an "alternation" (ABC/A'B'C'/A''B''C''/and so on). An example is Delilah's persuading Samson to reveal to her the secret of his strength in the Book of Judges (16:6–20). Four times she beseeches him (A, A', and so on); three times he "mocks" her (B, B', and so on); and three times he retains his strength (C, C', and so on). On the fourth occasion, however, he tells his secret (D, replacing B), and loses his strength (E, replacing C). The change in pattern not only brings us to a climax for which we have been prepared, it also marks the end of this particular section (or "panel") of the narrative. The whole of Judges 3–16 is structured around an alternation: namely, Israel Sinned (A), Israel fell into Servitude (B), Israel cried to the Lord (C), the Lord sent a Deliverer (D), and there was a Period of "rest" (E). This sequence occurs six times in the fourteen chapters (compare 2:11–19; see Parunak, "Oral Typesetting," 156–157). Obvious examples of the same pattern from our own folklore are the three journeys of Sir Bedivere to dispose of Excalibur (Malory, *Le Morte Darthur*, 21.5), and the three bears' discoveries of Goldilocks's interference in their domestic arrangements ("Who's been eating my porridge?" "Who's been sitting in my chair?" and so on).

Another pattern, particularly beloved of Hebrew tradition, involves concentric, bracketing, or ring composition, also known as "chiasm." Chiasm may have a double center (ABCC'B'A') or a single center (ABCB'A'). In its simplest form (ABA'), it is often called by the Latin name *inclusio*. There is a good example of chiasm in Genesis 3:9–19, where we are first told the sins of the man, the woman, and the serpent, in that order (A, B, C) (3:9–13); and then we hear the punishments of the serpent, the woman, and the man, with the order reversed (C', B', A') (3:14–19).

The effect of chiasm or *inclusio* is often to emphasise what is at its heart, as in David's poignant cry on learning of the death of his son:

> [A] O my son Absalom, my son, my son Absalom!
> [B] Would God I had died for thee,
> [A'] O Absalom, my son, my son!

This is an example of internal *inclusio*—that is, the bracketing elements are themselves part of the central statement. External *inclusio*—where the bracketing elements are outside the central statement—often serves to indicate material that is peripheral to the main action, though of interest. Thus, Solomon's correspondance with Huram of Tyre is preceded by a remark about those whom Solomon numbered (Hebrew, וַיִּסְפֹּר; RSV, "assigned") for the building of the Temple (2 Chron. 2:2 [Hebrew text, 2:1]); it is followed by another remark about how Solomon numbered (Hebrew, וַיִּסְפֹּר ; RSV, "took a census of") all the aliens living in the land of Israel (2:17 [Hebrew text, 2:16]). The effect is to recall us to the place we had reached in the narrative before the digression about the correspondence with Huram began. So used, *inclusio* is the simplest and most natural device in the world for a speaker, as those who try to tell stories, or even to give lectures, soon discover.

On occasion, similar features will occur at the end of one set of episodes and at the beginning of the next, so providing a "hook" that links the two "panels" together. Thus, Ezekiel 24 clearly consists of two quite separate parables about the destruction of Jerusalem by Nebuchadnezzar. The former is acted out by the prophet who boils a cauldron dry to symbolize the destruction of the city (24:3–14), the latter concerns the death of the prophet's wife (24:15–27). But at the end of the former parable, the prophetic word turns from speaking in the third person of the city to speaking in the second person to the prophet himself (24:13–14): this prepares us for the continuing address to the prophet that will dominate the second parable.

A particular episode may have features pointing both forward and backward, and so act as a "hinge" between two sections of narrative. Thus, Ezekiel 45:1–8 tells how the land is to be assigned to priests, people, and princes, in the reformed Israel. By its mention of priests and princes, however, the passage also serves as a hinge between the oracle preceding it, which describes the duties of priests (44:6–31), and that following it, which describes the duties of princes (45:9–25) (Parunak, "Oral Typesetting," 542).

Havelock prefers to speak of these arrangements not as "patterns" (a visual image) but as "acoustic responsions" ("Oral Composition," 183). Needless to say, the devices we have described do not function with perfect precision. They are creations of artistic instinct, not effects of analysis. Sometimes several of them will be functioning at once, so that a single passage demonstrates, say, both alternation *and* chiasm; or a single episode will function in two ways, as the concluding bracket of a concentric arrangement, and also as a hook into the next section of narrative. The basic principle is, however, always the same; it is, as Havelock points out, to

> avoid sheer surprise and novel invention, . . . the basic method for assisting the memory to retain a series of distinct meanings is to frame the first of them in a way which will suggest or forecast a later meaning which will recall the first without being identical with it. What is to be said and remembered later is cast in the form of an echo of something said already; the future is encoded in the present. All oral narrative is in structure continually both prophetic and retrospective. . . . Though the narrative syntax is paratactic—the basic conjunction being "and then," "and next,"—the narrative is not linear but turns back on itself in order to assist the memory to reach the end by having it anticipated somehow in the beginning ("Oral Composition," 183).

Oral Techniques and Communication

Oral techniques are valuable in orally composed poetry because they are tools for memory. Such is not only true for the poet; it is also true for the poet's hearers. Moreover, what the members of an audience can remember they may also be able to understand, and even to use for themselves. In other words, oral techniques, because they are tools of memory, are also tools for communication. As H. Van Dyke Parunak has shown in a number of articles, in an oral text the techniques of orality can play much of the part which punctuation, highlighting, headings, and paragraphs play in a graphic text ("Oral Typesetting,"

"Transitional Techniques"). We have already noted ways in which alternation and chiasm can be used to indicate that a section of the narrative has ended, or to bracket material that is peripheral to the main story. Such devices can also clarify shifts in thought, emphasize what is important, and offer refinements.

What happens, then, when the singer of tales takes up pen and papyrus, or begins to dictate? What happens when the bard no longer needs oral techniques as tools of memory? Actually, from the viewpoint of the audience, very little happens. For those who will *hear* the text, oral techniques continue to function as they have always done: they are still tools for communication and understanding. Thus, Virgil's repeated "pious Aeneas" (*pius Aeneas*) may not have for Virgil the memory-assisting function that "crafty Odysseus" had for Homer, but still it serves to render unforgettable that quality of Aeneas which characterizes him as a Roman rather than a Homeric hero, and which must be grasped if we are to understand the poem (for example, *Aeneid* 1.305, 4.393, 5.26, 286, 685). Similarly, when the exploits of heroes involve "standard themes," their stories are easier for us to follow: not to mention that the meeting of our expectations (balanced, no doubt, with enough variation to keep us interested) affords us considerable satisfaction.

Rhetoricians and critics of late antiquity were particularly aware of how useful for communication were the effects of oral structuring: alternations, concentric constructions, hooks, hinges, and so on. The narration of history, says Quintilian, requires not so much the full rhythms of periodic style (precisely what Mark lacks!) as "a certain continuity of motion and connection of style. All its members [*membra*] are closely linked together . . . ; we may compare its motion to that of people who link hands to steady their steps, and lend each other mutual support" (*Institutio Oratoria* 9.4.129–30, trans. H. E. Butler, adapted). Lucian and the writer of *Ad Herrenium* both comment on the importance of such structuring, "dwelling on the same topic and yet seeming to say something ever new. . . . We shall not repeat the same thing precisely—for that, to be sure, would weary the hearer and not refine the idea—but with changes . . . in the words, in the delivery, and in the treatment" (*Ad Herrenium* 4.42.54, trans. Harry Caplan). The "virtues proper to narrative," says Lucian, require that it progress "smoothly, evenly and consistently, free from humps and hollows." Clarity requires "the interweaving of the matter" so that, when narrators have completed one topic they will introduce a second, "fastened to it and linked with it like a chain, to avoid breaks and a multiplicity of disjointed narratives; no, always the first and second topics must not merely be neighbors but have common matter and overlap" (*How to Write History* 55, trans. K. Kilburn).

Oral techniques perform, then, the tasks of punctuation, clarifying, and emphasizing. From clarification and emphasis, it is only a step to commentary. Narration of an episode has an effect on us. When episodes are narrated in sequence, their effects interpenetrate each other. How much more terrible is the death of Hector (*Iliad* 22.136–404), because it follows so closely the account of his aged parents' prayers to him not to fight (22.25–92)—and how much more still if we remember the account of his farewell to Andromache in Book 6! Orality is not strong on the abstractions that commentary usually involves; but juxtaposition, repetition, anticipation, and bracketing often imply commentary. In this connection, reader-response critics have offered useful insights. Menakhem Perry speaks of the way in which "the order of a text creates its meaning" ("Literary Dynamics"). Shlomith Rimmon-Kenan points to what she calls "primacy" and "recency" effects: "Thus, placing an item at the beginning or at the end may radically change the process. . . . [B]oth the primacy and the recency effects may be so strong as to overshadow the meanings and attitudes which would have emerged from a full and consistent integration of the data of the text" (*Narrative Fiction*, 120–21). A stunning contemporary example of Rimmon-Kenan's point is offered by drama, a narrative form that Western societies still normally experience orally rather than graphically. In Shakespeare's *Hamlet*, the death of Hamlet is followed by the arrival of Fortinbras to restore order (5.2). So the play ends. The final presence of Fortinbras serves to remind us that despite all the chaos we have witnessed ("carnal, bloody, and unnatural acts"), still there is such a thing as the normal, and in the end it asserts itself. With that, we leave the theater. In Mel Gibson's filmed version of *Hamlet*, however, the arrival of Fortinbras is omitted. We witness carnal, bloody, and unnatural acts, and with those we leave the theater. What then? The omission of one short scene alters the effect of the entire action. Shakespeare's *Hamlet* presents us with chaos in rebellion against order; Mel Gibson's presents us merely with chaos. Such is the power of arrangement and "recency."

Oral Style and Mark

To what extent do we find in Mark the signs of oral style? This question will be discussed in the next six chapters. I propose in Chapter 9 to look at Mark's use of rhythm and formula, particularly in his arrangement of structure, and then in Chapters 10 and 11 at certain other details of his presentation.

BIBLIOGRAPHY

See the bibliography to Chapter 6. See also Walter J. Ong, S.J., *Orality and Literacy: The Technologizing of the Word* (London and New York: Methuen, 1982); and Eric A. Havelock, "Oral Composition in the *Oedipus Tyrranus* of Sophocles," *NLH* 16 (1984): 175–97.

On conceptual structural unity, see J. E. Grimes, *The Thread of Discourse* (The Hague: Mouton, 1975), 207–9. On the effect of structural arrangement as punctuation, see H. Van Dyke Parunak, "Oral Typesetting: Some Uses of Biblical Structure," *Biblica* 62 (1981): 153–68; and "Transitional Techniques in the Bible," *JBL* 102.4 (1983): 525–48.

On the effect of narrative arrangement on interpretation, see Menakhem Perry, "Literary Dynamics: How the Order of a Text Creates its Meaning," *Poetics Today* 1 (1979): 35–64, 311–61; Shlomith Rimmon-Kenan, *Narrative Fiction: Contemporary Poetics* (New York: Methuen, 1983); of course these matters are debated, and Perry and Rimmon-Kenan represent one side of an argument. For a window onto the debate, see Mikael C. Parsons, "Reading a Beginning/Beginning a Reading: Tracing Literary Theory on Narrative Openings," in Dennis E. Smith, ed., *How Gospels Begin*, Semeia 52 (Atlanta, Ga: Scholars, 1991), 11–32.

8

Mark and Oral Transmission

Understanding Mark's Structure

In Charles Williams's novel *Descent into Hell* there is an amusing moment when the hero Charles Stanhope, a poetic dramatist, reflects on the impossibility, or at least the unlikelihood, of explaining his work.

> There was a story, invented by himself, that *The Times* had once sent a representative to ask for explanations about a new play, and that Stanhope, in his efforts to explain it, had found after four hours that he had only succeeded in reading it straight through aloud: "Which," he maintained, "*was* the only way of explaining it."

That, in fact, is exactly what we must do if we are to appreciate Mark's structure. Attempts to analyze it in other ways—particularly by fitting it into the graphic logic of headings and subheadings—are numerous, and tend to differ from one another in striking ways (see, for example, Achtemier, *Mark*, 30–40; Dewey, "Mark as Interwoven Tapestry," 221–24). The reason for the problem is obvious: the logic of graphics does not coincide with the ways in which Markan structuring actually works. The most important elements in Mark's structuring are acoustic, precisely in accord with those characteristics of oral narrative style considered in Chapter 7. If we were to insist on representing Mark graphically, I suspect that the best way to do it would be by musical notation, with certain chords and rhythms representing particular themes and movements. Even this, however, would be an expedient. The only real way to understand Mark's structure is to follow it through as it was designed to be followed—indeed, to *listen* to it as it was designed to be heard—and so to experience it as it does its work.

Mark's Overall Arrangement

Even in orality, we have observed, chronology has some influence on narrative structure, simply because in narrative one thing happens after another. At an earlier stage of the discussion we noted some of the chronological elements in Mark. Therefore, despite my just-stated insistence that the most important elements in Mark's structuring are acoustic (and possibly as unwitting evidence that I am literate to my bones whether I like it or not!), I shall begin my examination by conceding that there is nonetheless something to be said for discerning an overall chronological arrangement, and that it constitutes a framework for the whole. Mark's "life" of Jesus appears to fall naturally into five parts, corresponding to five distinct movements in the action.

I. Prologue. Witness to the Coming One: In the Wilderness (1:1–8).
II. The Ministry of Jesus: In and around Galilee (1:9–8:21)
III. Jesus teaches the Way of the Cross: On the Road to Jerusalem (8:22–10:52)
IV. The Passion of Jesus: In and around Jerusalem (11:1–15:41)
V. Epilogue. Witness to the Crucified and Risen One: at the Tomb (15:42–16:8)

As indicated, these five parts also correspond broadly to rather striking changes of location; that we discern in our arrangement a certain rhetorical symmetry (compare Stock, *Method and Message of Mark*, 23–32) may further increase our confidence that what we think we are discerning is truly present. Indeed, I have little doubt that these five parts were broadly intended, and would broadly have been distinguished by early listeners.

The reasons for my precise delineation of the beginning and ending of each part will be indicated in what follows. However, as will also be shown, each part is linked to what precedes it and to what follows it by hinge passages (for example, 8:22–26). It is of the very nature of such passages that their relationship to what precedes and follows is ambiguous, if not ambivalent, just as a chord in music may be at once the resolution of one set of harmonies and the beginning of another. If, therefore, someone prefers, say, the more precisely geographic division of Bas M. F. van Iersel upon which mine is based (I. Wilderness—1:2–13; II. Galilee—1:14–8:26; III. The Way—8:27–10:52; IV. Jerusalem—11:1–15:41; V. The Tomb—15:42–16:8); or even the extended middle part of B. H. M. G. M. Standaert (6:14–10:52) upon whose arrangement van Iersel's is based, I shall not press the matter. Nor do I think such changes will affect many of the details in my "explanation" of Markan structure (see Standaert, *L'Evangile selon Marc*, 174; van Iersel, "De betekenis van Marcus," cited in Stock, "Hinge Transitions").

9

An Analysis of Mark's Structure

Part I. Prologue. Witness to the Coming One: In the Wilderness (1:1–8)

Mark moves us from our world to the world of his story by speaking at once of the "good news of Jesus Christ," which will be his central concern. The beginning of this good news was, he tells us, the coming of John the Baptizer in fulfilment of a divine promise made long before the plotted time of his story: "as it is written in Isaiah the prophet . . . John came" (1:2, 4). (I take it, following Basil the Great, Victor of Antioch, and probably Origen,[1] that 1:1–4 is a single sentence: a powerful rhetorical summons to attention that plunges us at once in medias res [Basil, *Against Eunomius* 2:15; this and other texts cited in C. H. Turner, "Marcan Usage," 145–46]). Thus, as Frank Kermode has pointed out, Mark's story implies and is part of a much longer story, which we already see stretching as far back as the prophets (Kermode, *Genesis of Secrecy*, 133–34).

The divine promise speaks of a voice "in the wilderness" says Mark (possibly somewhat misconstruing his original) (1:3), and it was "in the wilderness" (carefully repeated) that John spoke. "Wilderness" in Hebrew tradition, and perhaps universally, is always ambivalent. It is

1. Who were, after all, reading their own language, and were used to unpunctuated manuscripts.

a dangerous place, the abode of demons (Lev. 16:10; Tobit 8:3). It is also the place of rebirth and renewal: so it was for Israel at the time of the Exodus (Exod. 2:15); so it was for David (1 Sam. 23:14); so it was for Elijah (1 Kings 19:3–4); so it was in the traditions of Qumran (1 QS 8.13, 9.19); and so it was, says Mark, at the coming of Jesus.

Some things we learn at once from this opening experience "in the wilderness," and whatever confusions or complications the following narrative may bring, these first learnings will remain with us. First, we learn that Jesus is "the Christ," the "anointed": a word that, for all the varieties of understanding and hope that surrounded it, always had something to do with holiness, and with deliverance for God's people. Second, we learn that Jesus is "the Lord," whose "way" is to be prepared by God's messenger. Of these two things we are assured by the narrator (who, for the purposes of the narrative, we naturally assume to be omniscient) and by God. All else that we learn "in the wilderness" supports this: Jesus will be "stronger" than John the Baptist, and will baptize "with the Holy Spirit" (1:7–8).

So much regarding the immediate effect upon us of 1:1–8 may be said, I think, without controversy. Other aspects of the relationship of 1:1–8 to what follows and, in particular, to 1:9–15, are more doubtful, and certainly more debated. I reserve discussion of these questions for the next section.

Part II. The Ministry of Jesus: In and around Galilee (1:9–8:21)

Baptism, Testing, and Beginning of Ministry (1:9–15)

Like other heroes, Jesus is appointed for his task: at his baptism, the heavens are torn, the Spirit descends, and the heavenly voice declares him Son of God—the "beloved" son (1:10–11). Like other heroes, Jesus is tested: he is driven by the Spirit into the wilderness and tempted by Satan (1:12–13). Like other heroes, Jesus begins his task, proclaiming that God's "kingdom" (βασιλεία, transliterated, *basileia*, meaning "active sovereignty," or "kingly rule") is "at hand" and that *now* is the time for a change of heart and mind (repentance) and for trust in the good news (1:14–15).

Attempts to discern the exact relationship of the three episodes in 1:9–15 to what has gone before and to the rest of the gospel provide a good example of the problems of delineating Markan structure to which we referred in our previous chapter. Debate as to whether Mark's "introduction" extends only as far as 1:8, or to 1:11, or to 1:13, or to 1:15, has been extensive (see, for example, Boring, "Mark 1:1–15," 53–59, and literature there cited). In fact, a case can be made for all four views.

On the one hand, it was obvious that in 1:1–8 Mark was introducing us to his subject. He challenged us to move from our world to his, and established motifs (such as "Christ") that would be important in what was to follow. In 1:9–15, by contrast, he actually begins his account of Jesus' "life." The three episodes of 1:9–15 are, moreover, marked off from what has gone before by their own bracket (*inclusio*): Jesus comes from Galilee at the beginning of the first (1:9), and returns to Galilee at the beginning of the third (1:14). Bearing these factors in mind, we are bound to sympathize with those who regard 1:1–8 alone as Mark's "introduction."

On the other hand, the three episodes in verses 1:9–15 do have very close links with what precedes them. First, although Jesus appears "from Nazareth of Galilee" (1:9)—the early mention of Galilee at once pointing us forward to the main location of the next part of the narrative—nonetheless his baptism and testing are still located in the wilderness, and so are tied geographically to what has gone before. Second, John the Baptizer is still present: not only do we hear of his baptising Jesus (1:9), we are also told that Jesus' work does not begin until after John's arrest (1:14). Third, the whole of 1:1–15 is marked by an *inclusio*: Mark began by claiming to tell us "the beginning of the good news [*euaggelion*]" (1:1); and it is by "proclaiming the good news of God" that Jesus begins his ministry at 1:14.

Moreover, like verses 1–8, the three episodes in verses 9–15 establish motifs that will be important in what is to follow. Jesus as God's son (1:11; compare 9:7, 12:6, 15:39), Jesus tested by Satan (1:13; compare 8:33), and Jesus the messenger of God's kingdom (1:14; compare 4:11, 4:26–32, 9:1, 10:23, 11:10, 12:34, 14:25), are the most obvious. We ought also to note the reference to John's arrest, which is hardly careless or without art. In Mark's Greek, the arrest is spoken of as John's being "handed over," his παραδοθῆναι (transliterated *paradothēnai*) a form of the verb παραδίδωμι (transliterated, *paradidōmi*) (1:14).

Paradidōmi is a word that we shall hear again and again in Mark. Sometimes it is translated as "betray." Jesus, too, will be "handed over" (compare 3:19, 9:31, 10:33, 14:11, 14:18, 14:21). Coming at this stage in the narrative, *paradothēnai* tolls like a warning bell. We are in a world where *paradothēnai* describes the fate of God's prophets.

On grounds then both of its direct links with the first eight verses, and of the fact that like those verses it introduces us to what is to come, a case can be made by those who regard section 1:9–15 as part of Mark's introduction. With some changes of emphasis, cases can also be made for ending the introduction at 1:11 or 1:13. What then? The truth is, Mark's text is not really amenable to this kind of analysis. We should do better to judge Mark against the standard proposed by Lucian: "After the preface, long or short in proportion to its subject matter, let the transition to the narrative be gentle and easy" (*How to Write History* 55, trans. K. Kilburn). The three episodes in 1:9–15 are transitional, functioning as a kind of hinge, or series of hinges. Certainly they are closely tied to Mark's introduction, but they also begin his main narrative, and are filled with motifs pointing both forward and back.

Jesus Calls His Companions (*1:16–20*)

Heroes generally have companions—Achilles has Patroclus and his comrades-in-arms, Arthur his fellowship of the Round Table (Homer, *Iliad* 1.306–7; Malory *Morte Darthur* 3.2 [Vinaver, 1.98–99]). Sages, in particular, have (and often call) disciples (for example, Diogenes Laertius, *Lives of the Philosophers* 2.48, 7:2–3). So Jesus now summons those who are to follow him. Simon, Andrew, James, and John are names we shall meet again. Jesus has become the leader of a company, and will be so through all that follows, until his disciples foresake him (14:50).

A Day with Jesus (*1:21–39*)

"And they came to Capernaum" (1:21). Four episodes taken together form a day in the ministry of Jesus, from Sabbath morning to Sunday morning. He teaches with authority (1:22, 27); he casts out demons (he has already defeated Satan!) (1:25–26); he heals (1:29–30); he eats with his followers (1:31); he prays (1:35); and he goes to proclaim, for that is his purpose (1:38). (That, of course, is how his ministry was first shown to us, "proclaiming the kingdom of God" [1:15]). Crowds thronged to John (1:5), and now crowds throng to him (1:33). All this we might expect from the "anointed" who is "God's Son." Other aspects of the day are more difficult to understand, or more sinister. The demon calls Jesus "Holy one of God!" (1:24), and we know that the demon is right. Yet the demon is silenced. This is the first occasion when Jesus refuses to be spoken of in words that seem to be appropriate. It will not be the last. When Jesus goes out in the morning to pray, Peter and the others "hunted for him" (1:36) (κατεδίωξεν αὐτὸν, "followed hard upon him," "pursued him closely"—the verb is very strong). Peter wants to bring Jesus back to Capernaum, and comes as spokesman for popular enthusiasm: "Everyone is searching for you" (1:37). But Jesus must go on to other towns: that is his purpose (and, implicitly, God's purpose) (1:38). This is the first time Peter urges Jesus from a human viewpoint rather than God's; it will not be the last.

Cleansing a Leper (1:40–45)

Here is another hinge between two sections. Jesus' healing and his injunction to silence echo what has gone before (1:44). The instruction to do what "Moses commanded" (1:45) breaks new ground. It emphasizes that Jesus is a faithful Jew, and so creates an impression that will affect our attitude to those in the next section who will accuse him of flouting Jewish tradition. If we are listening to Codex Bezae, moreover, we find Jesus moved with *anger* in his confrontation with the leper (1:41). The reason is not entirely clear, but this, too, anticipates anger that will explode in the following sections.

Conflict with the Religious Authorities (2:1–3:6)

Five episodes continue to show Jesus as we are becoming used to seeing him: healing (2:1–12, 3:1–6), teaching (2:17, 2:18–22, 2:25–28), eating (2:16), calling a follower (2:14), accompanied by disciples (2:15; 2:18; 2:23), and accompanied by crowds (2:4). But now there is something new, something of which we have so far heard only hints (1:14, 1:41 Codex Bezae). There is conflict.

Joanna Dewey perceives in this section both linear development and concentric arrangement (*Markan Public Debate*). There is linear development in that the five episodes show Jesus in gradually mounting conflict with the establishment. In the first episode the conflict is muted and indirect: Jesus' critics merely question "in their hearts" (2:6) and seem at the end to join in the general praise (2:12). In the next episode they challenge his disciples, rather than him (2:16). In the next two episodes they challenge him directly (2:18, 2:24). In the final episode they actually begin by looking for a means to bring charges against him (3:2), and at its conclusion they are undisguisedly seeking "how to destroy him" (3:6). At this point we realize for the first time that their opposition is implacable. It will be some time before Mark says this so plainly again, but the point has been made. "Mark 2:1–3:6 might perhaps be compared in musical terms to the statement, early in a musical work, of a major theme which then hangs ominously over the composition, but which only comes to dominate the music much later in the piece" (Dewey, *Markan Public Debate*, 119).

The section also involves a concentric development. The question about fasting that forms its central episode (2:18–21) speaks of disciples who, because they are in company with Jesus, act in ways characteristic of the presence of the kingdom, a presence that explodes the boundaries of previous religious tradition (2:18–19, 21–22); at its heart this episode also speaks (for the first time in the gospel) of Jesus' coming death, and of disciples whose lives will go on (2:20). This central episode is bracketed by the call of Levi (2:13–17), and the disciples plucking grain on the Sabbath (2:23–28): two episodes that show the disciples enacting "kingdom behavior" (eating with sinners, acting as did David's followers) and, again, being justified for so doing because they are with Jesus. These in turn are bracketed by two outer episodes, the healing of the paralytic (2:1–11) and the man with the withered hand (3:1–6). By contrast with the inner episodes, neither of these stories directly involves the disciples, and both involve Jesus on a capital charge (blasphemy, Sabbath breaking). It will be on a charge of blasphemy that he will finally be condemned (14:64). Acoustically, then, we sense that what is for disciples the joy of the kingdom in Jesus' presence has at its heart and surrounding it death for him.

For all their variations in setting and ostensible subject, essentially all five conflict stories are about Jesus: who he is, what he does, and (above all) what is his authority. In hearing them we are learning a great deal about him: that he is Son of man who has authority to forgive sins (2:5–12); that he is physician of the sick, the one who calls sinners to repentence (2:17); that he is the bridegroom in whose presence fasting is improper (2:19–20); that he brings new cloth that cannot be used to patch old garments, new wine that cannot be put into old wineskins (2:21–22); that he is the one who may do as David did, the Son of man who is Lord of the Sabbath (2:25–28). (The text of the "plucking grain on the Sabbath" episode clearly centers on claiming Davidic authority for Jesus; it says nothing whatever about the disciples being hungry—a suggestion imported by Matthew [12:1–8] and Luke [6:1–5].) Twice, we note, Jesus calls himself "Son of man"—both times in connection with an authority (on the second occasion, a Davidic authority) that his critics are called upon to acknowledge. These are the first occasions when Jesus uses this title in the gospel; they will not be the last.

Finally we learn that although Jesus does what is lawful on the Sabbath, although he "does good" and "saves life," nevertheless he is the one whom various parts of the establishment in surprising alliance (the pious Pharisees with the worldly Herodians!) plan to destroy (3:4, 6). In the light of that knowledge we shall listen to all that follows.

At the Seaside (3:7–12)

A summary section reminds us of much that we have heard: the ongoing presence of Jesus' fishermen-disciples and of the crowds (3:7–9); Jesus' healing and his mastery of the demons (3:10–11); and his continuing (and still unexplained) demand for silence when the demons (correctly) name him, "Son of God" (3:11–12).

Polarizations, Parables, and Mighty Acts (3:13–6:13)

Continuing Mark's general description of Jesus' ministry, we are shortly to be presented with two sections of the gospel, each of which forms an identifiable collection: Jesus teaching by parables (4:1–34), and Jesus performing mighty acts (4:35–5:43). The latter follows the former without intervening material, but the two sections together are surrounded by a concentric arrangement. An inner pair of brackets consists of two episodes showing Jesus misunderstood and rejected by those whom we might loosely describe as "his own"—not merely the religous authorities (to whose opposition we are by now becoming used) but also his family and fellow citizens (3:21, 31–35, 6:1–6). An outer pair of brackets speaks of the Twelve (3:13–19, 6:6b–13). We first hear them called to exercise some part of Jesus' own ministry (3:14–15), and later we learn that this is what they have done (6:12–13).

Within this overall arrangement there are numerous other echoes and responses. Even as we rejoice in the call of the Twelve, we are reminded that one of them is Judas "who betrayed [παρέδωκεν] him" (3:19). In the following dispute with the religious authorities (3:20–29), Jesus, accused of casting out demons by satanic power, replies that it is only the one who has "bound" Satan who has power over Satan; and we, of course, know that the one who has "bound" Satan is Jesus himself. Moreover, we know that in winning that victory, he was driven by "the Spirit" (1:12). How dreadful then the act of those who declare the cause of his victory over the demons to be the work of an "unclean spirit"! (3:28–30). They are rejecting God's own work.

The following section on Jesus' family (3:31–35), while reinforcing the theme of Jesus rejected by "his own," is also reassuring for those who hear and identify with the gospel. Those who do the will of God "are my mother and my brothers and sisters!" (3:35). We have already noticed how naturally and easily these words could be used to include the listening audience as well as the characters in Mark's narrative. This assurance, in its turn, fittingly anticipates and introduces the collection of Jesus' parables (4:1–34)—a section that has aptly been called "The Great Assurance."[2] The parables themselves, as arranged, are calculated to assure the hearers—Jesus' first followers in the plotted narrative, and Mark's audience as the story is told—that despite all appearances and experiences to the contrary, despite all that will follow in the narrative and in the history of the church, despite the fact that some will fall away through persecution, the seed of God's word does bear its fruit, and the kingdom of God will triumph.

Within this overall arrangement, Mark introduces other themes. We are presented with the (characteristically rabbinic) pattern of public teaching for the crowd followed by private instruction for disciples (4:10–13), a picture that will be repeated (7:14–23, 9:14–29, 10:1–10, 13:1–8). We learn that there is a "mystery" (NRSV, "secret") of God's kingdom that already "has been given" to Jesus' disciples but which

2. The title was originally applied to the parables in Mark 4:1–34, and to other synoptic parables, by Joachim Jeremias in *The Parables of Jesus*: "out of the most insignificant beginnings, invisible to the human eye, God creates his mighty kingdom" (149; compare 146–53); compare also Stock, *Call to Discipleship*, 88–95—although Stock appears to me somewhat to have misread Jeremias.

"those outside" do not understand (4:11–12). But even as we are told this, we have for the first time a disturbing hint that the disciples themselves may not be so "inside" as they should be. Even they, apparently, do not really understand what Jesus is saying (4:13). Who, among Jesus' disciples or Mark's listeners, can answer with entire confidence the implied question in 4:15–20, "Are you good soil?" There is "nothing hidden except to be disclosed," but that does not relieve disciples of the need to "pay attention" to what they hear (4:24). These, too, are matters of which we shall hear again.

While I am not personally convinced by attempts to find an internal concentric structure in 4:1–34 (for example, Dewey, *Markan Public Debate*, 147–52; Donahue, *Gospel in Parable*, 30–32), the section as a whole is clearly identified by its own *inclusio*: the bracketing references to Jesus' extensive public teaching by parables (4:1–2, 4:33–36). Following the oral principle of development within the same, however, the second part of the *inclusio* also reiterates the newly introduced theme of private instruction to disciples following public teaching: "but privately to his own disciples he explained everything" (4:34b).

The immediately following section tells of Jesus' mighty acts, showing him master of the elements, the demons, disease, and death (4:35–5:43). The first mighty act is performed for male Jewish disciples; the second is for a gentile male on gentile soil; the third and the fourth (which are bracketed together) are for women—one, a woman with a hemorrhage; the other, a child. The woman with the hemorrhage is a particularly striking figure, and would have been especially so to those among Mark's hearers who were familiar with Jewish custom. As regards the cult, she is unclean: she has no business touching the holy man's hem, and she knows it (5:33). Mark also informs us that she has been economically exploited to the point where she has no more resources, and that she has been physically damaged (5:26): so her subterfuge arises out of her need. Jesus' response to her says nothing of uncleanness, nor of the boundaries of propriety. He calls her "Daughter" (5:34), a title that implies not exclusion, but family membership. We already know who are the members of Jesus' true family (compare 3:34–35); now we learn that a woman's presumption, breaking taboos against her sex and her condition, can be a manifestation of the faith that leads to that membership, and brings healing (5:34a). "Go in peace. Be whole from your disease" (5:34b). As Jesus is speaking, there is fresh news: "Your daughter is dead" (5:35). But Jesus is the conqueror of death, too (5:36–43), something that we shall do well to remember as Mark's story proceeds.

Throughout the accounts of the mighty acts, familiar themes continue to occur: the presence of the crowd (which forms, indeed, an important element in the plot of one episode [5:21, 24, 27, 30–31]), and Jesus' injunction to secrecy, which we are coming almost to take for granted, although we still do not know the reason for it (5:43). (Interestingly, the Gerasene demoniac, a gentile, *is* told to speak: but even he is not told to speak of Jesus; rather, he is to tell what "the Lord" has done for him, an appropriate enough instruction to a pagan. This is not to say that we may not also see a certain ambiguity in the instruction [5:19; compare 1:3]).

Taken alone, these vivid accounts of Jesus' power and compassion might have tempted us to suppose that we were faced in the gospel with a "superman" messiah who would proceed more or less effortlessly to triumph while we (our assumptions and attitudes largely unaffected) stood by and applauded. Such suppositions are at once given a sharp jolt by the account of the rejection of Jesus in his home town, Nazareth, and by his own people—the second element of Mark's inner bracket (6:1–6). "And they took offence at him" (6:3) becomes mere meanness in the face of all that Jesus has done and we, like Jesus, marvel at their unbelief (6:6). (The effect on us here is not unlike that in the closing sections of J. R. R. Tolkein's *Lord of the Rings*, where Frodo and his companions come home, having participated in the councils of the great and saved the world from tyranny, only to be jeered at by ignorant ruffians.) There is irony, moreover, in the reference to Jesus' mother and brothers and sisters, whom Jesus' critics claim to have "with" them (6:3): for we have been listening to Mark's story, and we know very well where Jesus' true family are to be found (compare 3:35, 5:34).

The hostility of Jesus' own is, however, only the inner bracket of the section. There is an outer bracket. "And he went among the villages teaching" (6:6b). So, finally, Mark brings us back to the Twelve (6:7–12), whom Jesus once more "calls to" him (προσκαλεῖται) (6:7; compare 3:13), and whom he now begins to "send out" (ἀποστέλλειν) (6:7; compare 3:14) to do the work for which he originally chose them: to preach, and to exorcise demons (6:7, 12; compare 3:14b-15). Jesus' teaching and Jesus' mighty acts may be surrounded by criticism and hostility from those who should have given him support, but surrounding that is something more important still: God's purpose remains, and the calling of those who are disciples does not change, whether their Lord is abused or not. "So they went out and preached that people everywhere should repent. And they cast out many demons, and anointed with oil many that were sick and healed them" (6:12–13). In the name of their Lord and in his power they share his work.

The Passion of the Baptist (6:14–29)

The episode of sending the Twelve not only brackets what has gone before: it also provides a hinge to the next section. Mark's concluding reference to the disciples' widespread preaching (6:12–13) prepares us for his next remark: "And King Herod heard" (6:14). In an instant, we have left the rural and urban scenes of Jesus' ministry, and we are in the royal palace.

Mark now does two things. First, he tells us something of rumors that are being offered as descriptions of Jesus: "John the baptizer has been raised from the dead." "It is Elijah." "It is a prophet, like one of the prophets of old" (6:14–15). This is the first time we have heard such rumors; it will not be the last. Second, Mark tells us how John the Baptizer died. The earlier mention of John's arrest tolled like a warning bell; now the same bell tolls again, louder and longer. We are in a world where God's prophets die at the hands of the wicked. The narratives of John's and Jesus' deaths resemble each other in overall content: each describes a just man done to death by a malicious and self-serving authority. They also resemble each other in structure. In each, one factor seems to favor the hero's survival: in John's case, Herod's superstition (6:20); in Jesus', the crowd's admiration (12:37b, 14:2). In both, the restraining factor is done away with: in John's case, by Herod's embarassment over his oaths (6:26), in Jesus', by the authorities' manipulation of the crowd (15:11). From that point, the end is inevitable. Both men die violently, and both accounts end with the body being laid in a tomb (6:29, 15:42–47).

Jesus Feeds His People and Rules the Waters (6:30–52)

Despite the death of a prophet, God's work continues. The Twelve return to Jesus (6:30–32). This episode in turn, however, enables the narrator to speak of the continuing pressure of the crowds (6:31–34), and so to move naturally to the feeding of the five thousand and Jesus' walking on the water (6:35–52).

More than one commentator has noted the Mosaic features of these episodes: Moses, too, fed his people in the wilderness, and Moses showed mastery of the waters. From the viewpoint of the gospel's structure, however, we note eucharistic features that appear here for the first time: Jesus takes, blesses, breaks, and gives (6:41). Other themes that have grown familiar continue to resonate: notably, the disciples' inability to understand Jesus, and the mystery that surrounds his person. These are shown with particular power in the episode of his walking on the sea. The scene is filled with mystery. "He intended to pass them by," says Mark, without explanation; by that reticence he leaves us with a deepened sense of one who is not understood (even by us!) and who is not at the disciples' disposal, just as he was not at their disposal earlier when they "hunted for him" on behalf of the crowds (compare 1:37). And yet, when they are afraid and cry out, he *is* at their disposal: "It is I; do not be afraid." When he is with them in the boat and the wind has ceased (as it did once before at his word [compare 4:39]), Mark speaks again of their incomprehension: "for they did not yet understand about the loaves, but their hearts were hardened" (6:52). Still the statement is enigmatic, obliging us to ask ourselves, Do we, Mark's audience, really "understand" about the loaves either? And what then of *our* hearts? And where now is Jesus in this narrative, if not increasingly isolated from everyone, even his disciples?

Mark's narration of this episode is a masterpiece of restraint, and in that restraint lies much of its power; analysis of its content, however, shows how much our understanding of its significance depends on our seeing its place in the gospel's structure.

Healing the Sick, the Dispute about Qorban, and Making All Foods Clean (*6:53–7:23*)

A narrative summary at 6:53–56 serves to bring Jesus back to Galilee, and reinforces other themes that are by now familiar: Jesus' healing presence, and the constant, and increasingly frantic, demands of the crowds. This prepares us to be the more struck by the attack made on him by the Pharisees over the question of ritual cleanness (7:2).

The actual question, "Why do your disciples not live in accordance with the tradition of the elders, but eat with hands defiled?" (7:5) does not seem, in spirit, to be very different from questions that have been put to Jesus before, such as the question about fasting ("The disciples of John and of the Pharisees fast, but your disciples do not fast" [2:18]), but on this occasion Mark takes the dispute in a new direction. Whatever historical realities may lie behind the confrontation here described, there is not the slightest doubt what it means in our narrative. Once more we have the pattern of public debate followed by private instruction for disciples, and here the significance is spelled out: "Thus he declared all foods clean" (7:19). Perhaps this was an important point for Mark to make to his audience (compare Rom. 14:1–4, 6). It is certainly perfect preparation for what is to follow.

A Ministry to Gentiles (7:24–8:10)

"And from there," says Mark, "he arose and went away to the region of Tyre [and Sidon]" (7:24). Gentile country! In case we have missed the point, Mark stresses it: a woman comes to Jesus for help, and she "was a Greek, a Syrophoenician by birth" (7:26). There follows an unforgettable exchange. Jesus' initial response to the woman reflects a fierce consciousness of Jewish privilege that modern commentators find disturbing. Perhaps the members of Mark's audience did not. In any case, the dramatic climax of the story does not lie there, but in the wit and grace of the woman's reply. While earlier commentators' talk of Jesus "enlarging his viewpoint" was inappropriate for this kind of narrative, still it is fair to point out that this is the one occasion in Mark where Jesus is shown losing an argument, and that it is to a woman and a foreigner. Naturally, her request is granted (7:29–30).

Mark continues with a string of not particularly accurate geographic terms (7:31), all clearly designed to stress that we are still in gentile territory. These introduce us to two further episodes. The first tells of a deaf-mute whom Jesus heals, so that those present (presumably gentiles) though charged (like Jesus' fellow Jews) to secrecy, cannot refrain from speaking of him and praising him (7:36–37). The second tells how, "in those days . . . a great crowd gathered" and "after three days" (the postresurrection language is unlikely to be accidental) Jesus fed them. The gentile feeding is virtually a repetition of the earlier Jewish feeding, and is clearly meant to be. Again we have the eucharistic notes: Jesus takes the loaves, gives thanks, breaks, and gives to the disciples to set before them (8:6). From small beginnings there is the same result: all are satisfied, and there is plenty to spare (8:8). In this setting, the disciples' slowness of understanding persists: "How can one feed these people with bread here in the desert?" they ask, as if they had seen nothing (8:4). In its setting, the moment is not without comedy: but through it, the ambiguity of their relationship to Jesus deepens. "And he sent them away; and immediately he got into the boat with his disciples, and went to the district of Dalmanutha" (8:10); that is, back to Galilee: back to Jewish soil, and his homeland.

The End of the Galilean Ministry (8:11–21)

Back to Jewish territory means, apparently, back to what Jesus experienced earlier—controversy with the Pharisees (8:11–13; compare 7:1–23). The Pharisees' request for a sign, brusquely rejected, means that controversy with this group brackets the entire gentile section. Again Jesus embarks with his disciples in the boat. Are we to encounter once more the pattern of private instruction for the inner circle following public teaching? Mark has established the pattern, and we might have expected it. In fact, we are met by what seems like an ironic parody: a conversation showing that Jesus' division from the disciples is scarcely less profound than his division from the Pharisees. The disciples have "only one loaf with them in the boat" (8:14b) and complain to one another that they "do not have loaves (ἄρτους)" (8:16, 17).[3]

The way in which this phrasing resonates with the disputes over table fellowship that troubled the early church can hardly be accidental. "Because there is *one* loaf, we who are many are *one* body," said Paul to his mostly gentile converts in Corinth (1 Cor. 10:17); and this, of course, was precisely what some among Jewish believers found hard to stomach (cf. Gal. 2:12–14). "We do not have *loaves*," complain the disciples: that is, we do not have the materials for separate tables and separate fellowships. The force of Jesus' response is clear. He alludes to Isaiah 6—the very passage from Isaiah earlier used to speak of those "outside" who have failed to understand the kingdom of God at all—"Having eyes do you not see, and having ears do you not hear?" (8:18; compare 4:11–12). Do the disciples not remember what happened when Jesus fed the crowds? Did not Jew and gentile alike receive the gift, and more than was needed? Twelve baskets more (the Israel number) for Jews? And seven more (the universal number) for gentiles? "Do you not yet understand?" (8:21).

3. This exactly translates Mark's Greek: contrast the RSV, which makes Mark write nonsense in view of 8:14b (Beck, "Reclaiming a Biblical Text").

Part III. Jesus Teaches the Way of the Cross: On the Road to Jerusalem (8:22–10:52)

The union of Jew and gentile in the gracious purposes of God goes to the heart of the gospel, and fittingly brings to a climax the first part of Mark's narrative. Yet even this does not finally address the mystery of Jesus' person. This mystery will dominate the third (and central) part of Mark's narrative, which is arranged around Jesus' journey with his disciples to Jerusalem.

The episode with which the section begins, Jesus' healing of the blind man at Bethsaida (8:22–26), functions as another hinge. It is a healing story, and is accompanied by the now-familiar request for secrecy: so it echoes themes with which we have become familiar. On the other hand, this is the first cure of blindness in the narrative, and to that extent it represents something new. Moreover, it is an unusual miracle in that Jesus seems unable to give the healing at once. The man's sight comes in stages. These two aspects of the miracle—Jesus seeking to give sight, and the slowness and difficulty of the process—anticipate two important themes of the third part of Mark's narrative.

Mark's third part proper begins with the episode at Caesarea Philippi, far from Jerusalem. "Who do people say that I am?" Jesus asks the disciples (8:27). We are prepared for their answers, for we have already heard popular views of Jesus circulating at King Herod's court: he is John the Baptiser, or Elijah, or one of the prophets (8:28; compare 6:14–15). "And you—who do you say that I am?" "You are the Messiah," says Peter (8:29). We know that Peter is right, for Mark has told us so from the beginning (1:1). There follows the immediate, and by now familiar, injunction to silence (8:30). But this time it is accompanied by an explanation: "The Son of man must suffer many things, and be rejected by the chief priests and the elders and the scribes, and afterwards be killed, and after three days rise again" (8:31). The juxtaposition of this assertion with the messianic claim is staggering in its effect. This at last is the reason for those injunctions to silence that have been puzzling us for so long. This is the key to the mysteries that have surrounded Jesus' person. Jesus must suffer, and after suffering he will rise again. In other words, even if we *do* know who Jesus is (as the demons do) we cannot speak properly of him, unless we will speak of his cross and resurrection. As Reginald H. Fuller succinctly expresses it, "For Mark, it is only the crucified One who is the Messiah" (*He that Cometh*, 23). The significance of this announcement is highlighted by Jesus' use of the title "Son of man." We last heard this title in controversy with his critics, as he claimed authority to forgive sins, and (Davidic) authority over the Sabbath (2:10, 2:28). There his authority was rejected; here it is asserted again before those who claim to have accepted it.

The uncertainty of this claim is at once emphasized by Peter's refusal to accept what Jesus says, and his willingness even to "rebuke" Jesus (8:32). But we are already prepared for the fact that disciples do not always understand what they are told (8:14–21); in particular, we are aware that Peter sometimes speaks on behalf of popular human opinion rather than for Jesus or God (compare 1:36–38). So we are not unprepared for the shattering rebuke that Peter now receives in his turn. "Get behind me, Satan! For you are setting your mind not on divine things, but on human things" (8:33). It is a terrible moment, the more terrible for the moment of triumph that has so closely preceded it. Not the Pharisees, nor the Sadducees, nor Caiaphas, nor Pilate, nor even Judas is ever identified with Satan. Only Peter, when he will not accept the cross. To refuse to accept a crucified messiah is to be identified with that very Satan whom Jesus has already defeated!

In a striking reversal of the "public teaching followed by private explanation" pattern, Jesus now summons the multitude with his disciples, and challenges all who will hear to consider what following him means. (The only "multitude" present is clearly Mark's audience, and from the viewpoint of a performer, the passage would function well delivered straight to it, much as actors deliver soliloquies in Elizabethan drama.) To follow the crucified is, says Jesus, to be prepared, in one's degree, to share the Crucifixion (8:34–37). For Mark's first hearers, particularly if they were *Roman* Christians, the cross involved in this call to follow Christ might well be a real cross, as their sisters and brothers had found a few years earlier.[4] When the threat of physical persecution was less immediate, no doubt the call assumed more metaphorical senses. Even so, it could never be a matter of mere asceticism—that "self-denial" that may as well as any luxury be the vehicle

4. Nero attempted to divert from himself suspicion of having caused the fire of Rome in 64 by convicting "vast numbers" (*multitudo ingens*) of Christians, many of whom were subsequently "fastened on crosses, and, when daylight failed were burned to serve as lamps by night" (Tacitus, *Annals* 15.44, trans. John Jackson).

of self-indulgence. What is involved here is the disciple's need and desire to be with Christ, whatever the cost; not a quest of suffering for its own sake, nor a passive acceptance of suffering, but willingness to suffer with Christ for the sake of the gospel (8:35).

Only thus shall one avoid being shamed before the Son of man "when he comes in the glory of his Father with the holy angels" (8:38). Again we have the title "Son of man" in assertion of authority, this time the final authority of the divine judgment. That understood, we see that suffering with Christ for the sake of the gospel is always with hope: "Truly, I say to you, there are some standing here who will not taste death until they see that the kingdom of God has come with power" (9:1). Those who will accept the crucified and who will suffer with him will know, even in this life, the sovereign power of God. This is the "mystery" or "secret" of God's reign which, even though they did not yet understand it, had from the beginning "been given" to those who chose to follow Jesus (compare 4:11).

Immediately following the claim made at Caesarea Philippi, and closely associated with it (the precision of the temporal link, "after six days" is, outside the Passion narrative, unique in Mark), we hear of the Transfiguration. Jesus is transformed before his disciples, his raiment "dazzling white as snow, such as no fuller on earth could bleach them." With him are Israel's heroes, "Elijah with Moses" (9:3–4). It is a splendid moment of glory, and will sustain us through much that is to come. But there is more. In the opening sections of his narrative Mark spoke of witnesses to Jesus: the Scriptures, John the Baptizer, and the heavenly voice (1:2–11). Now we are reminded of those witnesses. First, the heavenly voice, which repeats what it said at the baptism: Jesus is God's "beloved Son" (9:7; compare 1:11). Still there is to be secrecy about this, but now for the first time we learn when the secrecy will end. Jesus the Son of God cannot be understood or proclaimed apart from the cross and the Resurrection: therefore the disciples are forbidden to tell what they have seen, until the Son of man has risen from the dead (9:9). Then they may speak. Next, Mark reminds us of the witness of the Scriptures, and the Baptizer. "Elijah has come," says Jesus, "and they did to him whatever they pleased, as it is been written about him" (9:13). The reference to John, though oblique, is clear, and the reference to Scripture is explicit. All is being fulfilled, as we were assured from the beginning (1:2–3). Only here, too, there is now added the note of the cross: "they did to him whatever they pleased." We have heard the story of the Baptizer's death, so we know what this means. Christians at Rome are not the first to be called to suffer for their faithfulness, just as they will not be the last. The passion of John which foreshadows in form and substance the Passion of Jesus also foreshadows the passions of those who will follow Jesus.

There follows the story of the healing of the boy with an unclean spirit (9:14–29). The story does nothing to raise our sense of the disciples' competence, but makes us more aware than ever of Jesus' power, wisdom, and authority (9:18b-29). If the disciples would heal, as he does, then they must be persons of prayer, as he is (9:29; compare 1:35, 6:46). Again, we are made aware of Jesus' loneliness: "O faithless generation, how long am I to bear with you?" (9:19). That loneliness will become increasingly apparent as we travel with him the road to Jerusalem (compare 10:32).

The disciples and Jesus move south, "through Galilee" (9:30). Again Jesus speaks of his coming Passion, death, and Resurrection, in words that echo his previous prophecy (9:31). Again the disciples' response is inappropriate. The pattern of Caesarea Philippi is repeated—only worse. This time it is not Peter's concern with Jesus' destiny that is the problem, but the concern of all the disciples with their own preeminence. "Who among us is greatest?" they ask (9:33). "If any would be first they must be last of all and servant of all," says Jesus (9:35). How can it be otherwise among those who claim to follow the crucified?

There follows a series of episodes and sayings on relationship: to the vulnerable (represented by a child) (9:36–37), to eccentric or nonconforming disciples (9:38–41)[5], to "little ones" (9:42–49), to each other (9:50), to wives (the vulnerable partner in marriages of the period) (10:1–11), to children (again) (10:13–16), and to possessions (10:17–30). In all these passages the underlying emphasis, in vivid contrast to the disciples' concern as to who shall be "greatest," is on the strong *yielding* to the weak, the privileged *transferring* privilege to the underprivileged, the very wealthy *foregoing* the fruits of wealth for the sake of the gospel. It is striking that at the climax of this the disciples do seem, momentarily, to see the point. "Then who can be saved?" they ask. The answer, Jesus tells them, lies not in their attempts

5. In this particular narrative, the disciples' objection to one who is casting out demons in Jesus' name is doubly ironic in view of their own earlier failure to do the very same (9:38 compare 9:17–18).

at obedience, but in God for whom "all things are possible" (10:26–27). Is it then the case that those who attempt obedience are wasting their time? By no means: they will receive their recompense—with suffering! (10:28–30). The summary of it all is, "Many that are last will be first, and the first last"; in the context, a splendid paradox, threatening to those who seek or claim to be "greatest," yet full of promise for those who seek (but do not claim to be very good at) obedience.

"And they were on the road, going up to Jerusalem, and Jesus was walking ahead of them; and they were amazed, and those who followed were afraid" (10:32). So Mark vividly summarizes the closing stages of the journey south. And now for the third time Jesus speaks of his destiny, using a pattern of words becoming more and more familiar to us although, following the rhetorical principle of variation within the same, there is also development: "Behold, we are going up to Jerusalem; and the Son of man will be delivered to the chief priests and the scribes, and they will condemn him to death, and deliver him to the Gentiles; and they will mock him, and spit upon him, and scourge him, and kill him; and after three days he will rise" (10:33–34). The disciples' response is no better than on previous occasions. Indeed, their pretensions grow shriller. Our cheeks burn with embarrassment as we hear the agonizingly inappropriate request of James and John (10:37). No longer is it merely that they do not understand a crucified messiah; nor even that they wonder who is greatest among them; now they will have chief places in the kingdom. The ensuing dialogue makes plain the crassness of their misunderstanding, and (thereby) the increasing loneliness of their master:

> "You do not know what you are asking. Are you able to drink the cup that I drink . . . ?"
>
> "We are able."
>
> "The cup that I drink you will drink . . ." (10:38–39).

Jesus can only drive home the lesson again, contrasting what is to be their nature as a society with the world's understanding of community and authority, and linking the disciples' destiny with his, if they choose to follow him: "whoever would be first among you must be slave of all. For the Son of man also came not to be served but to serve, and to give his life as a ransom for many" (10:44–45; compare 9:35, 10:31).

In previous instances where Jesus has spoken of his coming death, he has spoken also of his resurrection: here he does not do so. The omission is important. What is at issue here is not Jesus' final vindication, but the disciples' utter misunderstanding, for they still think of the kingdom as a quest for personal glory. Therefore, what is emphasized is that the Son of man's being appointed "to serve" (διακονῆσαι) leads him to anything but personal glory. Far from being in a position to appoint others to his service, he must go to the service to which he has been appointed. That service is his death.

Thus the third part of Mark's narrative ends where it began, with Jesus destined to die, and with those who would be his followers challenged to share his obedience. The first must be last. At Jericho, he is near his journey's end. As he is leaving, accompanied by his disciples and the ever-present crowd, Mark tells us of one last episode on the way. It matches the first. Jesus heals blind Bartimaeus (10:46–52). For Mark, giving sight to the blind is the beginning and the end of Jesus' journey to Jerusalem.

Part IV. The Passion of Jesus: In and around Jerusalem (11:1–15:41)

The third part of Mark's gospel ends where it began, with references to Jesus' approaching death and to his giving sight to the blind. But, in agreement with the rhetorical principle that "we shall not repeat the same thing precisely," it does not end exactly where it began. At least two significant developments should be noted. First, at Caesarea Philippi we were told not merely that the Son of man would suffer, but that the Son of man "must (*dei*) suffer." In what sense, "must"? Were the words a veiled reference to God's purpose? And if so, why should the messiah's suffering be God's purpose? With Jesus' last words to the Twelve before they enter Jericho, Mark begins to suggest an answer to that question. The Son of man's death is not only his appointed service, it is also "a ransom for many" (10:45). This phrasing looks forward, anticipating something that we shall hear again in the Passion (compare 14:22–24).

Second, at Jericho blind Bartimaeus twice addresses Jesus with the explicitly messianic title "Son of David." On this occasion the title is not rejected or questioned, nor (in contrast to Caesarea Philippi) is there an injunction to silence. On the contrary, it is as the declared agent of David's kingdom that Jesus will enter Jerusalem (11:9–10). In short, references to Jesus' death and to his giving sight to the blind do not merely bracket the third part of Mark's narrative; they are also hinges, linking it to what is to come.

The fourth part of Mark, the story of Jesus in Jerusalem, falls into three obvious sections: Ministry in Jerusalem (11:1–12:37); the Farewell Discourse (12:38–13:37); and the Passion (14:1–15:47). It will be convenient to continue the examination along these lines.

Ministry in Jerusalem (11:1–12:37)

Jesus enters Jerusalem as the agent of God's kingdom (11:1–11).[6] His first significant public action is on the following day when he "cleanses" the Temple (11:11b, 11:15–19). Whatever the cleansing may originally have signified (and opinions differ) there seems little doubt as to what it signifies for Mark, since he has bracketed it with the episode of the barren fig tree (11:12–14, 20–25). That which pretends to fruit and has none is to be destroyed. The cleansing is a warning of imminent destruction.

Perhaps for Mark's audience, there was another nuance in the story as here told. The Temple was, says Jesus, to have been "a house of prayer for all the nations" (compare Isa. 56:7; Jer. 7:11; 1 Kings 8:41–43), but it has become "a lair for brigands" [σπήλαιον λῃστῶν; RSV, "a den of robbers"] (11:17). Would Mark's hearers have been familiar with the events of the Jewish War—in particular, with that version of events wherein the Temple in the days before its fall was regarded as in fact having been taken over by "brigands" (λῃσταὶ) (compare Josephus, *Jewish War* 4.138–50). It is possible.

6. Communities are not static. If (as is entirely possible) the Markan community first heard Mark's account of Jesus' entry to Jerusalem while the Jewish War was in progress, they will have been struck by many ironies; if (as is also entirely possible, especially if they were Roman) the same people listened to the account a year or so later, having watched Vespasian's triumphal entry into Rome in the spring of 71 as victor in the Jewish War, they will surely have been struck by an entirely new set of resonances and ironies. (For a reconstruction of Vespasian's triumph as it might have appeared to a contemporary, see Lindsey Davis's splendid thriller, *Silver Pigs*; for academic reflection on Mark's narrative in the light of such events, see Paul Brooks Duff, "The March of the Divine Warrior" [1993]).

Be that as it may, the end of the two interwoven episodes is that the disciples are encouraged to be themselves a house of prayer. "Have faith in God," says Jesus (11:22). Things impossible are possible for God, and so are possible for those who trust God (11:23; compare 9:29, 10:27). "Whatever you ask for in prayer, believe that you have received it, and it shall be yours" (11:24). One thing is required: the life of the community must be rooted in and guided by that compassion which it looks to receive (11:25).

The next two episodes remind us of earlier witnesses. At the beginning of Mark's story we heard testimony to Jesus from the Scriptures, John the Baptizer, and the heavenly voice (1:2–11). We were reminded of those witnesses at the Transfiguration, when we were still first wrestling with the dreadful prospect of a suffering messiah (9:2–13). Now, days before his Passion, we are reminded of them again. First, the witness of John. The authorities challenge Jesus. "By what authority are you doing these things?" He answers with a counterquestion: in effect, By what authority did John baptise? It is not politically expedient for them to answer, for they fear the people (11:32). But we who are listening know the answer, and we are reminded that John testified to Jesus (compare 1:7–8).

Second, the heavenly voice. "And he began to speak to them in parables" (12:1). As we listen to the story of the wicked husbandmen (12:1–12), all of us, if we have ever heard any Scripture at all, know that the vineyard is Israel and the owner is God (compare Isa. 5:1–7; Ps. 80:8–19). The owner sends many servants to the vineyard, and they are abused. At last he sends his "beloved son" (12:6) ("beloved" is probably Mark's own choice of word, as we can see by comparing the *Gospel of Thomas*'s version of the same parable [65–66]). The phrase resonates with us at once. We were at the baptism and we stood with the disciples on the Mount of Transfiguration. We know who the "beloved son" is (compare 1:11, 9:7).

Third, there are the Scriptures: "Have you not read this scripture: 'The very stone which the builders rejected has become the head of the corner . . . ?'" (12:10). Of course (on the principle that "we shall not repeat the same thing precisely") these episodes do not merely point back to the witnesses to Jesus. They also point forward. The wicked husbandmen took the beloved son, "and killed him, and cast him out of the vineyard" (12:8). The stone was "rejected" (12:10). We are reminded at once of Jesus' own prophecies of his destiny (8:31, 9:31, 10:33–34). We know what is to come.

The parable ends, and Mark tells us again of the authorities' hostility toward Jesus: they seek to "arrest" him. One thing restrains them. They "feared the multitude" (12:12). Once more, the ever-present crowd! The crowd has been a source of pressure upon Jesus in the past. Now it seems to be his protection and will be significant in what is to come. For the moment, Jesus' enemies withdraw (12:12c), but they do not thereby cease their hostility. On the contrary: "they sent to him some of the Pharisees and some of the Herodians, to entrap him in his talk" (12:13). The pious Pharisees and the worldly Herodians! It is precisely the curious and unlikely combination of which we last heard at 3:6. Thus reminded, we do not need to be told their object, for we already know it. It is to "destroy him." In pointing us back, Mark has again pointed us forward. He has also, in the short term, provided himself with an ideal hook whereby he may lead us directly to the next part of his narrative: the four questions that will be put, in turn, by Pharisees, Sadducees, the "good scribe," and Jesus himself (12:13–37).[7]

7. "The Bible four times ordains that children should be taught the significance of the exodus from Egypt. In the Haggadah of the Seder, the service of Passover eve, these passages are taken to allude to four different types of sons" (Daube, *New Testament and Rabbinic Judaism*, 163–64). Daube suggested (158–69) that the "four questions" to Jesus with which Mark now presents us originally stood as a version of the Four Questions asked by the four sons in the Seder:

1. questions of wisdom (*chokmah*), about points of law: correponding to the Pharisees' question about tribute (12:13–17);
2. mocking questions (*boruth*, "vulgarity"), usually about resurrection: corresponding to the Sadducees' question about the resurrection (12:18–27);
3. plain questions about the basis of piety and a good life (*derek 'rets*): corresponding to the question about the great commandment (12:28–34a);
4. a question about apparent contradictions in Torah, but having no reference to any point of law (that is, a question of *haggadah)*: corresponding to the question about the messiah as David's son and as his Lord (12:34b-37).

In the midrash of the three sons in the Passover Haggadah, moreover,

> only three of them ask, whereas in the case of the fourth his father must make the beginning.
> The Markan introduction to the fourth question, that by Jesus, is: "And after that no man durst ask him, and Jesus answered and said while he taught in the temple." This is an unmistakable reference to the Passover Haggadah which says, "And he who does not know how to ask, thou open (the instruction) for him." (Daube, 166–67)

This is an impressive set of correspondances, though I suspect that it tells us more about Mark's source than about Mark. It may point to an original setting for passion traditions in a Christian Seder.

The section appears to contain both a linear and a concentric development. Beginning with the subtle attempt of the Pharisees and Herodians to ensnare Jesus, we move through the cruder hostility of the Sadducees, to the friendship of the scribe (12:34), to the final teaching in which "none dared to ask him any question" although "the great throng heard him gladly" (12:37). In other words, while the section is certainly overshadowed by the opposition of the religious establishment to Jesus with which it begins, yet in its linear development, emphasis is not only on that opposition, but also on Jesus' triumph over it, and his "successful vindication of himself" at every point (Dewey, *Markan Public Debate*, 163).

On the other hand, we seem also to have here another example of concentric arrangement (compare Dewey, *Markan Public Debate*, 156–61). The four questions were introduced by the parable of the wicked husbandmen and the debate that followed it, that is, by an episode asking what it means to reject God's "beloved son" (12:1–12): they end with an episode centering upon the question, "Whose son is the messiah?" (12:35–37). Within these outer brackets, the inner pair of questions on tax to Caesar and the great commandment obviously centers on duty to God (12:13–17, 12:28–34). The central episode concerns God's faithfulness, in death as in life (12:18–27). So, viewed concentrically, the debates begin and end with the challenge of the person of Christ, while at their heart is the challenge to trust in God's faithfulness.

One overall effect of the narrative of the four questions is to sustain our awareness of the religious establishment's hostility toward Jesus, all the time diminishing that establishment and its representatives in our eyes. Thus the Pharisees (in league with the Herodians) appear as cunning schemers out to "trap" Jesus (the verb is ἀγρεύω, used of a hunter seeking prey) (12:13–17). The Sadducees are heretics who would mock him and the resurrection (12:18–27). And the scribes are those who simply do not understand who or what the messiah is (12:35–37). Even the dialogue with the "good scribe" does not much alter this impression, for now it is the scribe himself who renders the establishment relative, affirming that love of God and neighbor is "much more important than all whole burnt offerings and sacrifices" (12:28–34).

The four questions conclude the narrative of Jesus' public ministry in Jerusalem, which is marked, as we might expect, with its own *inclusio*. It began with Jesus' entry as Son of David, herald of David's kingdom (10:47, 48; 11:10). It ends with the affirmation that even that title, though not entirely wrong, is certainly not enough (12:35–37).

The Farewell Discourse (12:38–13:37)

The short section denouncing the scribes (12:38–40) is a hinge between Mark's account of Jesus' public ministry in Jerusalem, and the conversation and discourse with the disciples that is to follow. Criticism of establishment religiosity confirms the main themes of the former, while specific mention of those who "devour widows' houses" prepares us for the scene with which the latter will begin.

As Jesus watches the rich casting their gifts into the treasury, a widow puts in two small coins. He summons the disciples to hear his comment upon this generosity (12:41–43). For the last time in the gospel we move from public to private teaching. The story of the widow's mite is a good deal less straightforward than is usually admitted. The widow, says Jesus, has cast in "more" than all of them. In what sense, "more"? Clearly in the sense of commitment: the rest "have contributed out of their abundance; but she out of her poverty has put in everything that she had, all she had to live on" (12:44). So far we may understand the tale as exemplifying (and approving) total devotion and generosity, and so it is generally interpreted. But we cannot overlook the setting in which it has been placed. We have just heard Jesus denounce a religiosity that devours "widows' houses," and now we watch a widow giving up "all she had to live on" for the sake of the Temple. In a moment, moreover, we are to learn that of the "wonderful stones and wonderful buildings" to which she is contributing "not one stone will be left here upon another" (13:1, 2). Her devotion is wholehearted: but it is, apparently, to a cause that is futile (10:42–45). Hence the scene is rife with ambiguity. Certainly we may still feel admiration for the woman's generosity; but our admiration will be heavily conditioned by our questions about the doomed system that exploits her.

Jesus' words about the future of the Temple lead naturally to the disciples' next question: "When will this be?" which leads in turn to the extended section of teaching often referred to as the "Little Apocalypse" or the "Apocalyptic Discourse." In fact, save for verses 24 to 27, apocalyptic imagery is not its dominating characteristic. We have already noted its resemblance to the literary convention of peripatetic dialogue followed by seated conversation and discourse. In terms of Mark's structure it is perhaps more important now for us to note that this is Jesus' farewell.

Death is, undeniably, *the* threshold experience. To sit with a person or even an animal who is near death is to be acutely aware of being with them *in limine* (on the threshold). They are close to us, but soon they will go from us to a new sphere, and we shall not, at least in the immediately foreseeable future, be with them again.[8] The ancients were conscious of that threshold, as are we. More than once in biblical, Near Eastern, and Greco-Roman literature we see sages pausing "on the threshold" to give final exhortations and advice to their followers. Plato's account of the death of Socrates (*Phaedo*), and farewell speeches in the biblical literature such as those of Jacob and Joshua (Gen. 47:29–49:33; Josh. 23:1–24:30), are obvious examples. Among the writings of Mark's nearer contemporaries, we may cite Tacitus's account of the death of Seneca (*Annals* 15:62–63), (wherein the sage exhorts his friends and his wife to steadfastness that will enable them to accept their fates with serenity, as he does his); Seneca's own *Letters to Lucilius* (in which he prepares for death, guides his friend's moral progress, and affirms that such teaching as his does not die with the teacher [21.4–5]); 2 Timothy; and, of course, Lucian's splendid spoof of the whole thing, wherein Peregrinus, who desired "to benefit humankind by showing them the way in which one should despise death" (33), shortly before his demise "dispatched missives to almost all the famous cities—testamentary dispositions, so to speak, and exhortations and prescriptions—and he appointed a number of ambassadors for this purpose from among his comrades, styling them 'messengers from the dead' and 'underworld couriers'" (*The Passing of Peregrinus* 41, trans. A. M. Harmon). What all the cited examples have in common is the understanding that with the sage's death and departure, their friends and disciples face a difficult and uncertain future. The disintegration of the sage's vision is a real danger, and disciples will need to accept new and perhaps frightening responsibilities. Yet the sage's example will

8. Writing of his wife's death, C. S. Lewis described the experience with his customary accuracy: "How long, how tranquilly, how nourishingly we talked together that last night! And yet, not quite together. . . . I had my miseries, not hers; she had hers, not mine. The end of hers would be the coming-of-age of mine. We were setting out on different roads. This cold truth, this terrible traffic regulation ("You, Madam, to the right—you, Sir, to the left.") is just the beginning of the separation which is death itself" (*A Grief Observed*, 14, 15)

itself provide inspiration for facing the future. Needless to say, the inspiration and exhortation thus given is, from the viewpoint of the literature that enshrines it, directed not only to friends and disciples who appear in the narrative, or to the ostensible recipients of letters, but to all who may hear and choose to identify with the sage's vision.

Jesus' farewell discourse in Mark serves just such a purpose. It gives the disciples (and Mark's audience) information by which they may understand the coming persecution of the church (13:5–9, 11–13), the preaching of the gospel "to all nations" (13:10), the destruction of Jerusalem (13:14–20), the rise of false claimants to messiahship (13:21–22) (possibly another reference to events surrounding the siege of Jerusalem; compare Josephus, *Jewish War* 2.434, 442–44), and, "after that tribulation," may look for the final victorious presence of the Son of man (13:24–27). Thus the discourse looks beyond Mark's plotted narrative into the history of the church, the experience of Mark's own hearers, and beyond that to the end of time. We, if we choose to identify with the Lord's followers, are the ones finally exhorted to watch the signs (13:28–30), not to be disturbed because Jesus' words cannot fail, although the final date of the end is unknown to all save God (13:31–32), and, above all, to be alert (13:33–34). "Watch therefore—for you do not know when the master of the house will come, in the evening, or at midnight, or at cockcrow, or in the morning—lest he come suddenly and find you asleep" (13:35–36).

"What I say to you," says Jesus, "I say to all: Watch!" (13:37).[9]

The Passion (14:1–15:39)

Setting the Stage (14:1–11)

The opening to the Passion narrative centers upon the powerful story of a woman who anoints Jesus with costly ointment (14:3–9). This episode is bracketed (and so emphasized in its warmth and compas-

9. Greek ἀναγινώσκω (transliterated, *anaginōskō*) (like English "read") can mean either "read aloud" (e.g. Acts 8:30, Rev. 1:3, *Shepherd of Hermas* Vis. 1.3.3) or "read silently" (e.g. Aristophanes, *Knights* 117, Antiphanes, *Sappho* [Edmonds, *Fragments*, 2.262–65]); its exact sense in any particular instance can only be judged from context. Nevertheless, the enigmatic instruction at Mark 13:14, "Let the reader (ὁ ἀναγινώσκων) understand" is probably to be understood as a stage direction to "the one who reads aloud (that is, to the assembly)." If so, then in performance these words should be omitted, on the principle that one does not recite stage directions, one carries them out.

sion) by the machinations of those who will destroy him (14:1–2, 10–11). One thing gives the authorities pause, and with this we have grown familiar: it is the presence of the crowds (14:2; compare 12:12). In the anointing episode itself, the picture of Jesus "in the house" is by now familiar (14:3; compare, for example, 1:29, 7:24, 9:33), as is the presence of critical persons—presumably disciples—who think they understand what is going on, and don't (compare 8:32–33, 9:38–39, 10:13–14, 10:48–49). The observation that the woman has prepared Jesus' body for burial continues to reinforce our awareness of his coming death, and the promise of "a gospel" that will be preached "in the whole world" echoes a theme that was first heard in the farewell discourse (13:10). Early in the narrative we were told of a woman who "was ministering" to Jesus with his followers (1:31); now he is again cared for by a woman, although this time "some of those present" (presumably his followers) scold her (14:5).

The placing of the anointing narrative at this point means that the farewell discourse, which was preceded by a story of woman's offering, is also followed by one. In both stories the offering is extravagant. In both the object of generosity is under threat, and the generous gesture is apparently futile. Both are followed by prophecy. Yet the differences exceed the similarities. In the case of the anointing, there is no suggestion that the woman has left herself destitute, and Jesus' praise is not conditioned or ambiguous (14:3, 5; compare 12:44. Also 14:6–9; compare 12:43–44). The prophecy following the widow's mite spoke of doom; that following the anointing promises a lasting memorial: "wherever the good news is proclaimed in the whole world, what she has done will be told in remembrance of her" (14:9). With such notes of hope and danger, the stage for the Passion is set.

The Passion Narrative Proper (14:12–15:39)

Audiences, and therefore the creators of oral literature, tend to be conservative; that is to say, no narrative that is designed to be heard may depart very far from an accepted tradition. Opinions differ, but my own view is that so far as the overall structure of the Passion narrative was concerned, Mark was faced with a basic pattern that brooked little change: a pattern represented for us not only by Mark, but also by the (probably independent) *Gospel of Peter* (or what we have of it), and by the Gospel according to John. Something like

Last Supper
Gethsemane
Arrest
Trial before Sanhedrin
Trial before Pilate
Scourging
Crucifixion

will have constituted a basis too familiar to be avoided or drastically rearranged, even supposing a narrator wished to do so.

In considering this basis, we should note the point made by Albert Bates Lord ("The Gospels," 54–58), namely, that the Passion narrative has an overall form that resonates with the patterns of oral heroic narrative. Jesus' death is surrounded by motifs, such as betrayal by someone close, and the failure of friends and companions, that occur in the deaths of other heroes: Samson, Heracles, Arthur, and Roland, for example. Such heroes often in some sense die for others, and are always greater than the death that overtakes them.

Critics have long noted that the Passion and Resurrection narrative is the most closely articulated section of the gospel. "With compelling necessity and logic one event succeeds another," observed Karl Schmidt (*Der Rahmen der Geschichte Jesu*, 303). This is true. On the other hand, we must not overstate its significance. This is, after all, narrative. Claims by some critics that the Markan Passion narrative is "tightly plotted" certainly could not be accepted if we were to judge Mark by modern standards. Even comparison with a more sophisticated literary text of its own period, such as Plutarch's account of events surrounding the death of Caesar (*Caesar* 56.1–69.8), will oblige us to see the matter in perspective. Plutarch's account gives us something of cause and effect, what caused dislike of Caesar to mount, what restrained it, and so on. What Mark gives us is essentially an easily remembered sequence of vivid scenes:

The Priests' Plot
The Anointing by the Woman
The Treachery of Judas
Preparations for the Passover
The Last Supper—Prophecy of Betrayal

The Last Supper—"Institution"

The Last Supper—Prophecy of Denial

Gethsemane

The Arrest

The Trial before the Sanhedrin (Peter's Denial)

The Trial before Pilate

Mockery by the Soldiers

Journey to Calvary (Simon of Cyrene)

Crucifixion between Two Bandits

Mockery by Priests

Death watched by the Women

This is highly effective narrative, but it is narrative structured in the oral manner. By such a series of vivid scenes the *Iliad* describes Hector's death before Troy, and the events that accompanied it (*Iliad* 22). Moreover, there are within Mark's account a number of the kinds of inconsequentiality that orality tolerates easily, but over which "literate" minds have puzzled for centuries: What *exactly* did Judas tell the priests? Who was the young man who fled naked?

This granted, it is evident that Mark's passion story does, like the rest of his work, show signs of structuring, both in its internal arrangements, and in its relationship to other parts of the narrative. It is, nevertheless, structuring in the *oral* manner. First, we should consider the overall chronology. "You do not know," said Jesus in the farewell discourse, "when the master of the house will come, in the evening, or at midnight, or at cockcrow, or in the morning—lest he come suddenly and find you asleep" (13:35). Roman usage (originally military) divided the night into four three-hour watches, normally designated first watch, second watch, and so on (Jerome, *Letters* 140.8; Caesar, *Gallic War* 1.40; compare Mark 6:48). Mark shows Jesus using the popular equivalents. Still, the elaborate statement of fourfold possibility seems strange in the context of the farewell discourse. What is the reason for it? In fact, it seems deliberately to anticipate Mark's structuring of the first part of the Passion (R. H. Lightfoot, *Gospel Message*, 53). Thus:

"when it was *evening*" (14:17) Jesus came with the Twelve for the Last Supper;

at (presumably) about *midnight* Jesus "came and found the disciples sleeping" (14:40); and "immediately . . . Judas came, . . . and with him a crowd" (14:43), and the disciples fled (14:50);

when Jesus' trial before the Sanhedrin was coming to its conclusion, and as Peter below in the courtyard denied him the third time, "the *cock crowed* a second time" (14:66–72);

"and as soon as it was *morning* the chief priests, with the elders and scribes, and the whole council held a consultation; and they bound Jesus and led him away to Pilate" (15:1).

This precise match can hardly be accidental. Noticeably, moreover, the division of the narrative into three-hour periods continues, as Mark describes the remaining warfare of the Son of man. "It was the third hour, when they crucified him" (15:25); "when the sixth had come there was darkness over the whole land until the ninth hour" (15:33); and "at the ninth hour Jesus cried with a loud voice, . . . 'My God, my God, why hast thou forsaken me?' . . . uttered a loud cry, and breathed his last" (15:34–37). As R. H. Lightfoot observed, such precision is the more striking in a writer for whom temporal precision is not usually a concern (*Gospel Message*, 53). It might well, originally, have been a way of remembering the story. As matters stand, its effect on an audience is undoubtedly to give this part of Mark's narrative a suggestion of clipped, almost military order. Indeed, it is noticeable that as the soldiers take over in the final stages of the action, so Mark's temporal language conforms to a semimilitary usage, giving the Crucifixion something of the feel of a report from the front. It is very easy for a listener to follow.

So much for overall chronology. What of details of the narrative? At the Last Supper, Jesus takes, blesses, breaks, and gives, echoing what he has done before (14:22–26; compare 6:30–43, 8:1–10). Yet, as always, it is not quite as before, because now he interprets what he does. The disciples earlier complained when they had "only one loaf" (8:14). Now Jesus tells them what the fellowship of those who share the one loaf means: union with him—"This is my body" (14:22). On the road, before Jericho, he had spoken of being appointed to serve by giving his life as "a ransom for many" (10:45). Now those words, too, are echoed and amplified. The shared cup is his "blood of the covenant, which is poured out for many" (14:24). Mark is not a theologian of the atonement, but he leaves us in no doubt that what is to happen on the cross will not merely be a result of human malice, nor even of Jesus' willingness to endure human malice, but "for many"—and hence, if we are willing to hear it, "for us."

So the Last Supper narrative echoes and amplifies themes that have preceded it. It also looks forward. Jesus' prophetic words anticipate his own final isolation, the disciples' failure, Peter's denial, and the cross (14:27, 29–31). They also anticipate victory: Jesus will next drink wine "new in the kingdom of God" (14:25). "I will go before you to Galilee" (14:28). That Jesus speaks of all these matters in the course of the supper strengthens our impression that the entire ensuing narrative is bound together and, indeed, that in all that happens, despite being victim, Jesus is also in the deepest sense in control.

The account of Jesus' arrest (14:43–50) is bracketed by episodes emphasizing the disciples' weakness and incomprehension (14:32–42, 14:51–52). The account of his trial before the Sanhedrin (14:55–65) is similarly bracketed by the episode of Peter's denial, with its own explicit reminder of Jesus' prophecy at the Last Supper, now fulfilled (14:53–54, 14:66–72). Understandably, these elements have led some critics to speak of Mark's hostility toward the disciples and toward Peter. In fact, there is no hostility in the narrative toward any of the disciples, and the effect of performance would, I suspect, be to lead most of us to identify with their failure rather than to scorn it. In any case, the failure of companions is a part of the normal machinery of oral heroic narrative: it serves to remind us the hero's greatness.

In the section before the arrest we hear not only of the disciples' weakness, but also of Jesus' prayer. After the cleansing of the Temple we heard him tell the disciples, "Whatever you ask for in prayer, believe that you have received it, and it will be yours" (11:24). It would be idle to pretend that such an exhortation did not involve problems of theodicy, of which the ancients were as well aware as we. Lucretius spoke scornfully of those who

> . . . revert again to ancient faiths,
> And for themselves take cruel masters who
> (For so the wretches reckon) can do all:
> They know not what can be and what cannot . . .
> (*De Rerum Natura* 5:86–87)

In the present passage Mark addresses such a view directly. Jesus' prayer begins by asserting a principle he has earlier presented as the only basis whereby we may hope for salvation at all: that all things are possible for God (compare 10:26–27) (precisely the principle denied by Lucretius). So Jesus prays: "Abba, Father, for you all things are possible." He continues, "Remove this cup from me; yet, not what I want, but what you want" (14:36). Trusting that God *is* his faithful Abba, and that God's will *will* be done, Jesus embraces that will, even if it involves his own death. As we listen to Jesus' prayer, of course we know that he is right to believe that God is his faithful Abba, for we have heard the heavenly voice (compare 1:1–3, 11, 9:7). So knowing, we are naturally encouraged to pray ourselves as Jesus prays: we are to pray, as Sharyn Echols Dowd powerfully expresses it, "expecting power and accepting suffering" (*Prayer*, 33).

During Jesus' trial, some speak of the destruction of the Temple: "We heard him say, 'I will destroy this temple that is made with hands . . .'" (14:58). If we recall the farewell discourse, we know that they are lying. Indeed, Mark explicitly tells us that they bore "false witness" (14:57). Jesus did not say that he would destroy the Temple; but he did say that it was doomed (13:2). So the words of Jesus' enemies indirectly remind us of that doom, and of doom for the system that the Temple represents. The accusers are already judged. The scene is replete with irony.

The high priest questions Jesus, and in so doing not only raises the same question as was raised at Caesarea Philippi, but also reminds us of God's own declaration at the Baptism and the Transfiguration: "Are you the Christ, the Son of the Blessed?" (14:61). Now, for the first and only time in the narrative, and at the moment when we cannot doubt what this will mean for him, Jesus' answer is explicit: "I am." Moreover, in words that echo both Caesarea Philippi and the farewell discourse, he continues to look beyond the cross, "You will see the Son of man sitting at the right hand of power, and coming with the clouds of heaven" (14:62; compare 8:38, 13:26). As in his first public responses to his critics, he claimed for himself the authority of the Son of man in forgiveness and in lordship over the Sabbath (2:10, 28), so now in his last public words he claims the authority of the Son of man in judgment. Such a claim might seem like madness: yet when, a few minutes later, we return to the courtyard and to the fulfilment of his prophecy regarding

Peter's denial, this merely confirms our impression that he is the one who is in charge here, depite all appearances to the contrary. Those who presume to judge him understand nothing. The irony deepens.

"And as soon as it was morning the chief priests with the elders and scribes, and the whole council held a consultation; and they bound Jesus and led him away and delivered him to Pilate" (15:1; compare 10:33!). An important thread binds Jesus' trial before Pilate to the rest of Mark's narrative. It is the presence of the crowds. This has been a feature of the story from the beginning, and twice recently we have been told that their support alone restrains the authorities from acting against Jesus (12:12, 14:2). So now they have their chance to play a part. "Do you want me to release for you the King of the Jews?" (15:9; compare 15:12). But crowds can be manipulated, and the authorities have taught them to prefer Barabbas (15:11). As for Jesus, "Crucify him!" they cry (15:13). As noted earlier, the structure of Jesus' Passion echoes the structure of John the Baptist's passion: with the restraining factor removed, Jesus' doom is certain (15:15).

A second thread binds together the narratives of Jesus' trial before Pilate, his being mocked, and his Crucifixion. It is kingship. Jesus is referred to as king no fewer than six times in thirty-one verses, and he is called so by virtually everyone involved: by Pilate (three times; the third time Pilate admits that he is quoting the crowd) (15:2, 9, 12–13), by the Roman soldiers (15:18), by the official *titulum* over his cross (15:26), and by the chief priests and scribes, who also speak of him as "the Christ" (15:32). Of course, the titles are given in mockery; but what for characters in the story is mockery, is for Mark's audience, the truth. Jesus *is* the Christ, the King of Israel. Hence irony surrounds the entire proceeding. All these people (like the bystanders who criticized the woman who anointed Jesus [14:4]) think they understand what is going on. But they don't.

The final act of Jesus' ministry is his death. It is told with simple dignity. Jesus' last cry, "My God, my God, why hast thou forsaken me!" (echoed in the original Aramaic, as if we dare not forget even its syllables) seals the loneliness that has been growing around him from the beginning. Structurally, however, we should notice four things about Mark's account of Jesus' death. First, there is a reference to Elijah, for the onlookers think that Jesus is calling him (15:35–36); at the first act of Jesus' ministry, his baptism, Elijah was also symbolically present, in the person of John. Second, Jesus "breathed his last [ἐξέπνευσεν, transliterated, *exepneusen* expired]" (15:37); at Jesus' baptism, the Spirit (πυεῦμα, transliterated, *pneuma*) descended upon him (1:10). Third, the curtain of the temple was torn (ἐσχίσθη, transliterated, *eschisthē*) in two, from top to bottom (15:38); at Jesus' baptism he saw "the heavens torn (σχιζομένους, transliterated, *schizomenous*)" (1:10). (Josephus tells us that the outer veil of the temple was a tapestry that "typified the universe" and that "portrayed on this tapestry was a panorama of the entire heavens" [*Jewish War* 5.212–14]). Fourth, and finally, the centurion declares, "Truly this man was the Son of God" (16:39); the baptism, too, ended with the declaration that Jesus was God's son (1:11). In short, Mark begins and ends his portrayal of Jesus' ministry with what David Ulansey aptly describes as a "cosmic *inclusio*" ("Heavenly Veil Torn," 123). We end where we began. Yet (as always!) not quite. To begin with, two of the verbal echoes we have noticed are just that, merely verbal echoes. The onlookers who think that Jesus is "calling for Elijah" are wrong. They think they know what is happening, but they don't. Similarly, the word with which we hear of Jesus' death (*exepneusen*) does remind us that the Spirit (*pneuma*) descended into him at his baptism; but the fact remains that the word itself means simply "died," as does our English word "expired." Mark's text does not tempt us to become docetists. More important, while the rending of the Temple veil may indeed remind us of the rending of the heavens at Jesus' baptism, still, Mark has told us far too much of the future of the Temple for us not to see in that sundering also a sign of coming destruction of the Temple itself.[10] Most important of all: once only God could declare the divine sonship; now a gentile soldier, who has seen and accepted the scandal of the cross, can do the same.

It is all watched by the faithful women. The passion narrative began with a woman's service to Jesus, and it ends with a group of women who watch. Mark lists the chief of them carefully by name, and adds that there were "many" others (15:40–41). Their presence adds poignance to the male disciples' absence. It also points forward. The drama is not yet over.

10. That Christians should link the destruction of the Temple, with the rejection of Jesus was natural enough. Possibly they were not the only Jewish group to see a connection between the destruction of the Sanctuary and the rejection and death of a messianic claimant; see Hengel, *The Zealots*, 296.

Part V. Epilogue. Witness to the Crucifed and Risen One: At the Tomb (15:42–16:8)

Of the Resurrection narrative, as of the Passion, we begin by noting that it resonates strongly with the structures of myth. Invariably, we have observed, mythic heroes are greater than the deaths that overtake them. Of Samson it is merely said that "the dead whom he slew at his death were more than those whom he had slain during his life" (Judg. 16:30). In other cases the victory is made explicit: Heracles became a god, Roland was taken up to heaven by angels, and with Arthur, *rex quondam et futurus*, there always lingers the hope that he will return. So in the Markan narrative Jesus has been betrayed by one disciple, forsaken by all the rest, mocked, crucified and entombed for the sake of others by his enemies; now he is vindicated and declared victorious.

What of the details of Mark's arrangement? The episode describing Jesus' burial is closely linked to what has gone before. It is linked temporally ("when evening had come"); it is linked by the appearance of some of the same characters—Pilate, the centurion, and the faithful women who watch—and it is linked by its triple statement that Jesus is "dead" (15:44–45). It introduces us briefly to Joseph of Arimathea, and he is the agent who brings us to the tomb.

Still the faithful women are present. They alone have the courage to honor Jesus in death. We began this story in the wilderness, a place of new beginnings but also a dangerous place, whence came a message of life. We end at the tomb, a place of death that turns out also to be a place of new beginnings, for from it comes another message of life. We began with a herald, and we end with a herald. We began with witness to Jesus the Coming One. We end with witness to "Jesus of Nazareth who was crucified. He is risen" (16:7). The elaborate *inclusio* looks back but also forward. The moment has come to which the disciples were pointed after the Transfiguration: the Son of man has risen from the dead (9:9). Then they were told to be silent; now is the time for them to tell what they have seen. "Go, tell his disciples and Peter that he is going before you to Galilee; there you will see him, as he told you" (16:6). What then? The women flee. They are seized by terror. They say nothing to anyone, for they are afraid (16:8).

At this point, Mark leaves matters. The unknown rhetorician whom classical tradition calls "Demetrius" (who probably flourished at some time during the first century B.C.E. or C.E.) would perhaps have approved: "not every point should be punctiliously treated with full details, but some should be left for the hearers to comprehend and infer for themselves" (*On Style*, 222). We may be fairly sure that the evangelist had never heard of "Demetrius," but his instinct was good. The ending to the gospel is rhetorically sound, and enables it to achieve a remarkable effect. As Reginald H. Fuller has pointed out, "Mark is not concerned with past history but rather with present proclamation. What has ongoing, contemporary relevance is not the original foundation of the church, but the extension of the church through the apostolic mission" (*Formation of the Resurrection Narratives*, 68). So, as the women leave the tomb, we, Mark's listeners, identify with them. We, too, know what it is to have seen a vision, to have been called to witness, and to be afraid. So what shall we do? Fearing Caesar, shall we deny the crucified and risen One? Or shall we confess? We know what the women did. We know what even the disciples did, incompetent failures though they have been throughout most of the gospel story. But what shall *we* do? Mark can say no more. That question we must answer for ourselves. So the evangelist returns us from his narrative world to our world, and to the future.

(The so-called longer ending to Mark [16:9–20] had been added by the end of the second century [Irenaeus, *Against all Heresies*, 3.10. 6; compare possibly Justin, *Apology*, 1.452]. It was probably a result more of theological concerns than of literary. It might have been good enough for "Demetrius" to leave some points for listeners to "infer for themselves," but it was not so, we may suspect, for many who held themselves responsible for the teachings of the Church.)

BIBLIOGRAPHY

The earliest substantial treatment of a gospel in the light of orality of which I am aware is Charles H. Lohr, "Oral Techniques in the Gospel of Matthew," *CBQ* 23.4 (1961): 403–35. Lohr's summary and illustration of oral narrative techniques is still valuable. Werner H. Kelber, *The Oral and the Written Gospel: The Hermeneutics of Speaking and Writing in the Synoptic Tradition, Mark, Paul, and Q* (Philadelphia: Fortress, 1983) was a pioneering attempt to apply insights from the study of orality to Mark. For discussions of some of

Kelber's positions, see Lou H. Silberman, ed., *Orality, Aurality, and Biblical Narrative* (Decatur, Ga: Scholars, 1987), especially the papers by Thomas J. Farrell, S.J., ("Kelber's Breakthrough" [27–46]) and Thomas E. Boomershine ("Peter's Denial as Polemic or Confession: The Implications of Media Criticism for Biblical Hermeneutics" [47–68]). Kelber's work remains, of course, a source of valuable insight. In my view, however, he underestimated the degree of continuity between oral and manuscript culture. Much of what he says about literacy (for example, 90–131) is far more applicable to print literacy then to the literacy of the first century (and even in that context, in the light of later work such as Zumthor's, would appear to be overstated). For a useful discussion of this aspect of Kelber's work, see Mary Ann Tolbert, *Sowing the Gospel: Mark's World in Literary-Historical Perspective* (Minneapolis: Fortress, 1989), 44–45, note 36. For further discussion of Kelber's positions, see Chapter 10.

On Mark's narrative structure, see generally: Norman R. Peterson, *Literary Criticism for New Testament Critics* (Philadelphia: Fortress, 1978); Joanna Dewey, *Markan Public Debate: Literary Technique, Concentric Structure, and Theology in Mark 2:1–3–6* Dissertation Series 48 (Chico, Calif.: Scholars, 1980); Augustine Stock, O.S.B., *The Method and Message of Mark* (Wilmington, Del.: Glazier, 1989), and *Call to Discipleship: A Literary Study of Mark's Gospel* (Wilmington, Del.: Glazier, 1982); Paul J. Achtemeier, *Mark* (Philadelphia: Fortress, 1986). There are many useful insights about Markan structure in Robert M. Fowler, *Let the Reader Understand: Reader-Response Criticism and the Gospel of Mark* (Minneapolis: Fortress, 1991), although Fowler would not himself concede that this is the subject of his study. See also Joanna Dewey, "Oral Methods of Structuring Narrative in Mark," *Interpretation* 43.1 (1989): 32–44; "Mark as Interwoven Tapestry: Forecasts and Echoes for a Listening Audience," *CBQ* 53.2 (1991): 221–36; Augustine Stock, "Hinge Transitions in Mark's Gospel," *Biblical Theology Bulletin* 15 (1985): 27–31; *The Method and Message of Mark* (Wilmington, Del.: Glazier, 1989), 19–32. For the comparison of Mark's structure with music, note Howard Clark Kee, *Community of the New Age: Studies in Mark's Gospel* (Philadelphia: Westminster, 1977), 64, 75.

On Mark's opening, see M. Eugene Boring, "Mark 1:1–15 and the Beginning of the Gospel," in Dennis E. Smith, ed., *How Gospels Begin*, Semeia 52 (Atlanta, Ga: Scholars, 1991), 43–81. Boring makes important points and has many valuable observations; he also cites most of the relevant literature. On the problem of punctuating 1:1–8, however, I remain convinced, against Boring, by C. H. Turner, "Marcan Usage: Notes, Critical and Exegetical, on the Second Gospel," *JTS* 26 (1925): 145–46.

On Markan usage of the phrases "kingdom of God" and "Son of man" see further Chapter 11 and literature there cited.

On the call narratives in the gospels, there is useful material in Arthur J. Droge, "Call Stories in Greek Biography and the Gospels," SBL 1983 Seminar Papers, 244–57. Ed. Kent Harold Richards Scholars Press, Chicago: 1983.

On 2:1–3:6, see especially Joanna Dewey, *Markan Public Debate: Literary Technique, Concentric Structure, and Theology in Mark 2:1–3:6*, Dissertation Series 48 (Chico, Calif.: Scholars, 1980).

On 4:1–25, there are useful insights in John R. Donahue, S.J., *The Gospel in Parable: Metaphor, Narrative, and Theology in the Synoptic Gospels* (Fortress: Philadelphia, 1988), 28–52; Thomas E. Boomershine, "Epistemology at the Turn of the Ages in Paul, Jesus and Mark: Rhetoric and Dialectic in Apocalyptic and the New Testament," in Joel Marcus and Marion L. Soards, eds., *Apocalyptic and the New Testament* (Sheffield: JSOT, 1989), 159–63. For discussion of the parables in narrative context, and especially Mark 4, see John Paul Heil, "Reader-Response and the Narrative Context of the Parables about Growing Seed in Mark 4:1–34," *CBQ* 54.2 (1992) 271–86 and literature there cited.

On 5:24b-34, and on the subject of women in Mark generally, see Marla J. Selvidge, *Woman, Cult, and Miracle Recital: A Redactional Critical Investigation on Mark 5:24–34* (Lewisburg: Bucknell University; London and Toronto: Associated University Presses, 1990). On Levitical laws regarding the menstruating woman, see Jacob Neusner, *The Idea of Purity in Ancient Judaism* (Leiden: Brill, 1973).

On the feeding stories, see Robert M. Fowler, "Loaves and Fishes: The Function of the Feeding Stories in the Gospel of Mark," *SBLDS* 54 (Chico, Calif.: Scholars, 1981).

On 8:14–21, see Norman A. Beck, "Reclaiming a Biblical Text: The Mark 8:14–21 Discussion about Bread in the Boat," *CBQ* 43 (1981): 49–56.

On 8:26–10:52, see Ernest Best, *Following Jesus: Discipleship in the Gospel of Mark* (Sheffield: JSOT, 1981), 15–146; Donald Senior, C.P., *The Passion of Jesus in the Gospel of Mark* (Wilmington, Del.: Glazier, 1984), 28–36; Augustine Stock, O.S.B., *The Method and Message of Mark* (Wilmington, Del.: Glazier, 1989), 230–87. On Mark 10:45, see John N. Collins's magnificent study, *Diakonia: Re-interpreting the Ancient Sources* (New York: Oxford University Press, 1990).

On mythic patterns in the gospel narratives, see Albert Bates Lord, "The Gospels as Oral Traditional Literature" in William O. Walker, ed., *The Relationships among the Gospels: An Interdisciplinary Dialogue* (San Antonio, Tex.: Trinity University Press, 1978) 33–91. On orality and the hero, see Gregory Nagy, *The Best of the Achaeans: Concepts of the Hero in Archaic Greek Poetry* (Baltimore, Md.: Johns Hopkins University Press, 1979).

On Mark 11:22–25 and 14:35–36, and the relationship of these passages to the rest of Mark, see the excellent study by Sharyn Echols Dowd, *Prayer,*

Power, and the Problem of Suffering: Mark 11:22–25 in the Context of Markan Theology (Atlanta, Ga: Scholars, 1988).

On Mark 12:41–44, see Addison G. Wright, S.S., "The Widow's Mite: Praise or Lament?—A Matter of Context," *CBQ* 44.2 (1982): 256–65; for sensible qualifications to Wright, see Elizabeth Struthers Malbon, "The Poor Widow in Mark and her Poor Rich Readers," *CBQ* 53.4 (1991): 589–604.

On Jesus' farewell discourse, see Vernon K. Robbins, *Jesus the Teacher: A Socio-Rhetorical Interpretation of Mark* (Philadelphia: Fortress, 1984), 171–79; also Leo G. Perdue, "The Death of the Sage and Moral Exhortation: From Ancient Near Eastern Instructions to Graeco-Roman Paraenesis," in Leo G. Perdue and John G. Gammie, eds., *Paraenesis: Act and Form*. Semeia 50 (Atlanta, Ga: Scholars, 1990), 81–109. See also Stock, *The Method and Message of Mark* (Wilmington, Del.: Glazier, 1989), 321–48. On possible "false messiahs" and the Jewish Revolt of 66–70, see Richard A. Horsley and John S. Hanson, *Bandits, Prophets, and Messiahs: Popular Movements in the Time of Jesus* (Minneapolis, Minn.: Winston, 1985), 118–27; Martin Hengel, *The Zealots: Investigations into the Jewish Freedom Movement in the Period from Herod I until 70 A.D.* (Edinburgh: T & T Clark, 1989), 293–98.

For the view of the pre-Markan Passion narrative taken here, see Helmut Koester, *Ancient Christian Gospels: Their History and Development* (London: SCM; Philadelphia: Trinity, 1990), 216–40; Joel B. Green, *The Death of Jesus: Tradition and Interpretation in the Passion Narrative* (WUNT 2/33; Tübingen: Mohr [Siebeck], 1988); Adela Yarbro Collins, "The Composition of the Passion Narrative in Mark" *STR* 36.1 (1992) (57–77). Obviously, this view differs sharply from that of, for example, Werner H. Kelber: on which, see the bibliography to Chapter 10.

On the chronology of the Markan Passion narrative, see Augustine Stock, *Call to Discipleship* A Literary Study of Mark's Gospel (Wilmington, Del.: Glazier, 1982), 177–79; also R. H. Lightfoot, *The Gospel Message of St Mark* 2d ed., corrected (London: Oxford University Press [1950] 1952), 51–53.

On Mark's alleged hostility to Peter, see Thomas E. Boomershine, "Peter's Denial as Polemic or Confession: The Implications of Media Criticism for Biblical Hermeneutics," in Lou H. Silberman, ed., *Orality, Aurality, and the Biblical Narrative* (Decatur, Ga: Scholars, 1987). On Mark's attitude to the disciples generally, see C. Clifton Black, *The Disciples According to Mark: Markan Redaction in Current Debate* (Sheffield: JSOT 1989), chapters 3–5. Bertram L. Melbourne, *Slow to Understand: The Disciples in Synoptic Perspective* (New York: University Press of America, 1988) contains useful material (including a review of earlier work on the subject [1–23]) but is somewhat convoluted by the writer's preoccupation with Markan nonprimacy.

For further suggestions regarding Markan irony in the accusation that Jesus threatened to destroy the Temple, see Donald Juel, *Messiah and Temple: The*

Trial of Jesus in the Gospel of Mark, SBL Dissertation Series (Missoula, Mont: Scholars, 1977).

On the Markan *inclusio* at the beginning and the end of Jesus' ministry, see David Ulansey, "The Heavenly Veil Torn: Mark's Cosmic *Inclusio*," *JBL* 110.1 (1991): 123–25, and literature there cited.

On Mark's ending, see Reginald H. Fuller, *The Formation of the Resurrection Narratives* (New York: Macmillan; London: Collier-Macmillan, 1971), 50–68; Thomas E. Boomershine and Gilbert L. Bartholomew, "The Narrative Technique of Mark 16:8," *JBL* 100.2 (1981): 213–23; Thomas E. Boomershine, "Mark 16:8 and the Apostolic Commission," *JBL* 100.2 (1981): 225–39.

10

Oral Characteristics of Mark's Style

Oral Style

We have examined the ways in which Mark's structure is typical of works styled for oral transmission. In our preliminary comments on oral style, however, we noticed a number of oral characteristics other than structural. We commented on orality's penchant for narrative rather than proposition, for the concrete and visual rather than the abstract, for hyperbole rather than balanced observation, and for confrontation rather than dialogue. Let us now look at some of Mark's material in the light of these considerations.

Episodes Showing Jesus' Wit and Wisdom

Many Markan episodes show Jesus in discussion or debate with disciples or opponents. The form of such episodes has been compared with rabbinic disputations and with Hellenistic *chreiai*. (A χρεία [transliterated *chreia*]—literally, "basic need" or "requirement"—is described by the grammarian Aelius Theon [c. 50–150 C.E.] as "a concise and pointed account of something said or done, attributed to some particular person" [*Progymnasmata*: *Peri Chreias* (Hock and O'Neil, 82–83)]).

Neither the rabbinic nor the classical comparison is inapt, and we have at least one early testimony that Mark's materials as a whole were perceived as "formulated for *chreiai*":

> Mark became Peter's interpreter, and wrote accurately all that he remembered, not, indeed, in order, of the things said or done by the Lord. For he had not heard the Lord, nor had he followed him, but later, as I said, followed Peter, who used to give teaching formulated into *chreiai* (πρὸς τὰς χρείας) but not making, as it were, an arrangement of the Lord's oracles, so that Mark did nothing wrong in thus writing down the single points as he remembered them. (Papias, c. 125 C.E.; quoted in Eusebius, *Church History* 3.39.15, trans. Kirsopp Lake, altered)

What is important about such comparisons from our viewpoint is not that the gospel materials are thereby bound into one particular cultural milieu, whether Hellenistic or Jewish, but rather that all three types of material—Hellenistic, Jewish, and that from the gospel—show to a greater or lesser extent the signs of oral composition.

Markan episodes showing Jesus in discussion or debate often begin with a question. There are occasions when the question is not overtly hostile (for example, 10:17–22—the rich man's question about eternal life; or 12:28–34—the scribe's question about the great commandment); and sometimes it is implied rather than explicit (for example, 3:31–34—the implied question being, Who are the members of Jesus' true family?); but generally the question is explicit and antagonistic, and it may even be accompanied by the observation that it was intended to test or entrap Jesus (as at 10:2 and 12:13). The questioners generally make or imply an opinion that is directly contrary to what the story will actually teach. They suggest, for example, that to heal on the Sabbath must show that one cares nothing for the Sabbath (3:2), or that belief in the resurrection of the dead is absurd (12:18–23). As far as the structure of the narrative is concerned, naturally the whole point of the question is that it leads to Jesus' memorable saying—which will answer it. Because of the question and the questioners—and this is especially so when the question is hostile—the issue is polarized and dramatized, and we, the listeners, are involved in it. Such an involvement prepares us to hear, remember, and, because we are *involved*, to claim for ourselves Jesus' words. Oral memory thrives on polarities. After hearing the story of the man with a withered hand whom Jesus healed on the Sabbath (3:1–6), of course we can see that compassion overrides the Sabbath. After hearing the story of Jesus' clash with the Sadducees over belief in the resurrection (12:18–26), of course we know that the dead are raised.

Settings to such stories, where they are given at all, are little more than sketches: a party in a tax collector's house, for example (2:15). But perhaps for that reason they are the more easily envisioned. Being envisioned, they provide a further link between our memories and the lesson to be learned. Bare assertion of Jesus' concern for sinners could easily be forgotten; reasoned argument might lose our attention; but we are not likely to forget the scene of that meal, and that incisive (and rhythmic) declaration,

> Those who are well have no need of a physician,
> but those who are sick:
> I came not to call the righteous,
> but sinners to repentance (2:17).

> Οὐ χρείαν ἔχουσιν οἱ ἰσχύοντες ἰατροῦ
> ἀλλ' οἱ κακῶς ἔχοντες·
> οὐκ ἦλθον καλέσαι δικαίους
> ἀλλὰ ἁμαρτωλούς.

This is the oral way of teaching: not by discussion of principle, but by vivid presentation of precept.

Episodes Showing Jesus the Man of Deed

Many episodes in Mark focus on Jesus the "man of deed" who performs mighty acts of healing and power. Here too we find the characteristics of oral narrative. The stories have simple and easily remembered plots, and follow a common pattern containing two main elements: Jesus, and the obstacle that needs to be overcome, such as the demon-possession, the storm, or the disease. Jesus is always heroic: the strong Son of God, mighty in deed. His motives for acting are generally assumed, but not discussed. Once, before healing a leper, he is said to be "moved with compassion" (1:41) (or, according to Codex D, "moved with anger"—an interesting variant in view of orality's penchant for confrontation); he is also described as "moved with compassion" (6:34) before he begins to teach the Five Thousand, so perhaps in Mark's view that observation could be said to cover the whole section. But these are the exceptions. Those in need of help (the diseased, the demon-possessed) hardly emerge as characters in their own right, and third parties in the narratives (such as the crowd, the disciples, or the Pharisees) function generally as suppliants (2:3–5, 7:32b), chorus (2:12b, 4:41, 5:17, 5:31, 5:35, 5:38, 6:51, 7:36–37, 9:28–29) or occasionally as part of the problem (2:6, 3:2). They, too, rarely emerge as individuals (though occasionally we gain a vivid glimpse of character through action, as with the Syro-Phoenician woman, or the father of the epileptic boy). The miracle stories, as has often been pointed out, sometimes contain vivid detail: but its function is not in general to make the characters real to us. Its function is most often to emphasize the difficulty of the miracle, and hence to increase our admiration for it (4:37–38, 5:2–5, 5:25–26, 5:35, 5:38–40, 8:2–3, 8:23–24, 9:17–24).

Parables

Mark presents Jesus himself as an oral teacher (4:33–34), and examples of this are scattered across the gospel. The major tool here, as Mark points out (4:33–34), is the parable. Parables speak of one thing in terms of another. (Greek *parabolē*, like Hebrew *mashal*, means literally "comparison"). Often parables speak of something that we might find difficult or painful to grasp in terms of something familiar and easy—a person lighting a lamp or a doctor visiting the sick (4:21, 2:17). Because parables operate in this way, two things follow. First, they are open-ended, always requiring audience response to intepret and complete them. Second, they are flexible in application. Hearing the Parable of the Sower, at one time we may identify with the soils ("What kind of soil am I?" is the question implied by 4:15–20) and at another with the sower ("Will there be any results from all this work?"). For each of us, therefore, the effect of the parable is slightly different. In short, the meanings of parables do not lie merely in the analogies they offer or (in the case of narrative parables) in the stories themselves, but in the significance we give them in different contexts. The parable's evident expectation of audience response, and its flexibility, naturally denote it as suited to oral communication. Understandably, Werner Kelber speaks of parables as "a quintessential *oral form of speech*" (his italics) (*The Oral and the Written Gospel*, 62).

The Markan parables show various internal signs of oral patterning.

Thus the Sower (4:3–9) and the Wicked Husbandmen (12:1–10) offer classic examples of repetition of sequence ("alternation"). Four times the sower casts seed (A, A', etc.); three times it falls on unsatisfactory soil (B, B', etc.); three times it fails to produce a crop (C,C', etc.). On the fourth occasion it falls on good soil (D replacing B), and is successful (E replacing C). Four times the owner of the vineyard sends to the vineyard for his share of the produce (A,A', etc.); three times he sends a servant or servants (B,B', etc.); three times the husbandmen ill-treat or kill them (C,C', etc.). On the fourth occasion the owner sends his son (D replacing B) and the husbandmen kill him to claim the inheritance for themselves (E replacing C). In each case, the change in pattern brings us to the climax for which preceding elements in the story have prepared us: in the case of the sower, the abundant harvest; in the case of the husbandmen, judgment.

Parables, we have observed, speak of what is unfamiliar or difficult in terms of what is more familiar. That in itself does not help memorization, since what is familiar—the ordinary—is scarcely what is memorable. However, in Markan parables what is ordinary is often highlighted by being contrasted with what is fantastic or absurd, or by the indication of something extraordinary about it. Someone lights a lamp and puts it onto a lampstand, in contrast to the absurd notion of lighting a lamp and placing it under a measure or a bed (4:21). (It is notable that it is the absurd part of the parable which, in its King James form, has entered common English parlance, even among those who, one suspects, do not know precisely what "bushel" means.) A farmer sowing seed is ordinary enough: but consider how the crop from the seed on good ground is out of all proportion to what is sown! (4:2–9) A seed grows; we all know that. But consider the mystery: it grows by itself, and we don't know how! (4:26–29) Consider how huge is the mustard plant in comparison to its tiny seed! (4:30–32) No doubt the picture of angry tenants and a demanding landowner was familiar enough in first-century Palestine: but this landowner, who seems at first to be a reasonable enough fellow, takes measures to obtain his share of the crop that move from the reasonable to the absurd (12:1–12). Other typical events—the fig tree putting forth its leaves as the first sign of summer (13:28), or servants waiting for their master to return (13:34–36)—might be familiar enough, yet so important in the lives of Jesus' hearers as to need no exaggeration or emphasis.

Parables, as said, invite our participation: "To you has been given the secret of the kingdom of God, but for those outside everything is in parables: so that they may indeed see but not perceive, and may indeed hear but not understand; lest they should turn again, and be forgiven" (4:11–12; compare Isa. 6:9–10). The key to the meaning of this passage is in its allusion to Isaiah 6, which speaks of the *pathos* of God who sends prophets that we may hear, and is continually ignored. Parables invite our participation, but it is not merely a participation of the mind that they seek. "If we do not obey we are rebels." The world of Mark's parables—the world of lampstands and measures and seeds growing secretly (4:21–29)—is above all a world in which we must *act.* Lamps must be set on lampstands; measure must be given as well as taken; the harvest must be gathered. The landlord's servant is coming for his produce (12:1–11); the master of the house is coming home (13:34–36). What then shall we do? Mark's parables, like his teachings on prayer, invite us to accept what befalls us, and also to act in hope, in ways that befit the kingdom. Mark's overall purpose, we have observed, involves epideictic with deliberative elements: that is, the rhetoric of the gospel is calculated to persuade those who hear it to particular attitudes, and to particular kinds of action in the future. The parables are a vivid and effective tool for that purpose. Those who will receive the mystery of the kingdom, they imply, are those who will attempt obedience.

Narrative Summaries

From time to time in Mark's narrative we find short, generalizing summaries and descriptions (what Karl Ludwig Schmidt [*Der Rahmen der Geschichte Jesu*] called *Sammelberichte*).

> And he went throughout all Galilee, preaching in their synagogues and casting out demons (1:39)
>
> He went out again beside the sea; and all the crowd gathered about him, and he taught them (2:13)
>
> Jesus withdrew with his disciples to the sea, and a great multitude from Galilee followed; also from Judea and Jerusalem and Idumea and from beyond the Jordan and from about Tyre and Sidon a great multitude,

> hearing all that he did, came to him. And he told his disciples to have a boat ready for him because of the crowd, lest they should crush him; for he had healed many so that all who had diseases pressed upon him to touch him. And whenever the unclean spirits beheld him, they fell down before him and cried out, "You are the Son of God." And he strictly ordered them not to make him known (3:7–12).

Other episodes listed by Schmidt include 1:14–15 (a summary of the Galilean Ministry), 1:21–22 (Jesus at Capernaum, teaching with authority), 1:39 (Jesus' tour of Galilean synagogues), 2:13 (Jesus by the sea), 4:33–34 (parabolic teaching), 6:7,12–13 (the mission of the Twelve), 6:30 (the return of the Twelve), and 6:53,56 (the landing at Gennesaret). Here too, as Joanna Dewey has recently pointed out ("Oral Methods," 36), we find the style of oral narrative. Scenes such as those of general healing in 3:7–12 and 6:56 are straightforward, highly concrete, and easily visualized. They stand in relationship to other units recounting specific episodes rather as, in the *Iliad*, general descriptions of the battle stand to the accounts of individual combats.

The Passion

Irony in the Passion Narrative

Mark's passion narrative is, we have observed, replete with irony. There is irony in the false accusations of those who say that Jesus has doomed the Temple, never dreaming that in a sense they are right. There is irony in the mockery of those who call him king, not knowing that he is. There is irony in the paradox, "He saves others; he cannot save himself" (15:31). There is irony in the presence of faithful women at cross and tomb, whence male disciples, chosen and sworn-to-faithfulness, have fled.[1] All this is the irony beloved of orality, which delights in ugly ducklings who are really swans and frogs who are really princes.

1. This irony was not lost on the seventeenth-century English bishop Lancelot Andrewes, who seized on it joyfully in an Easter Sermon:

> . . . as it was a special honour . . . so was it, withal, not without some kind of enthwiting [reproach] to them, to the Apostles, for sitting at home so drooping in a corner, that Christ not finding any of them is fain to seek Him a new Apostle; and finding her [Mary Magdalen] where He should have found them and did not, to send by the hand of her that he first found at the sepulchre's side, and to make himself a new Apostle (*Ninety-Six Sermons*, 5.44).

With just such irony we hear the wicked stepsisters mock Cinderella, never dreaming that it is for her foot that the glass slipper was made. With just such irony we hear Penelope's evil suitors mock Odysseus, never dreaming that he is master of the house, and that their doom is sealed (*Odyssey* 18–22).

Death in the Passion Narrative

It seems hardly necessary to say that Mark's passion narrative is concerned with death. There is no attempt to shield us from the horror of death, either by noble sentiment or theological reflection. Jesus himself is "greatly distressed and troubled." "My soul is very sorrowful" (14:33, 34). The Crucifixion is narrated with brief, blunt realism. Indeed, as we have observed, Mark seems at times to emulate the detachment of a military report. Nothing in the narrative allows us to deceive ourselves about what is happening, up to and including the final cry of desolation, "My God, my God, why hast thou forsaken me?" (15:34)—words so deeply burned into the tradition, it seems, that they are remembered in the original Aramaic, even though the rest of the story has long been told in Greek.

Werner H. Kelber has spoken of "orality's reluctance to speak of death" (*The Oral and the Written Gospel*, 194). I must respectfully disagree. Such concern with death and such realism in speaking of it is entirely characteristic of orality. It is present in *Beowulf* and *La Chanson de Roland*. As for Homer, I am reminded of a remark made by Gregory Nagy in his masterly study of the hero in archaic Greek poetry, *The Best of the Achaeans*:

> I wish only to insist on the most fundamental aspect: that the hero must experience death. The hero's death is the theme that gives him his power. . . . Not even the lofty Olympians can match that, since they cannot die; when the pro-Achaean gods enter combat with their pro-Trojan counterparts in *Iliad* XXI, the results cannot be fatal—and they cannot be serious either (9, sec. 17).

Life in the Passion Narrative

Of course we must not distort or exaggerate this concern with death, either in oral literature generally, or in Mark. We have just spoken of Mark's irony. We have noted the extent to which the hero of oral heroic narrative is in some sense victor in death: and Mark's closing chapters, certainly, are not merely about passion and death, but also about passion and resurrection. At each key point in the narrative the most memorable word is given not to death but to life. At the Last Supper, Jesus says that he will not drink wine again, but there is a promise even in that: he will not drink, "until that day when I drink it new in the Kingdom of God" (14:25). In the same scene there is a warning to the disciples, who (in the general manner of the companions of heroes) have failed often and will fail again, "You will all fall away"; but much more important is Jesus' promise, "But after I am raised up, I will go before you to Galilee" (14:27–29), a promise whose fulfillment will be triumphantly picked up in the final revelation at the tomb. At the trial we see Jesus humiliated and silent, yet the most powerful word in the scene is when Jesus breaks his silence in response to the direct challenge, "Are you the Christ, the Son of the Blessed?" "I am; and you will see the Son of man seated at the right hand of Power, and coming with the clouds of heaven" (14:61–62). At the cross, of course, there is Jesus' death, but in that very instant there is also the centurion's testimony, first fruit of the gentiles: "Truly this man was God's Son!" (15:39).

As it happens (but it is surely no accident) most of the elements we have been identifying have important functions in the structure of Mark's narrative. And they *all* point to Jesus' vindication. What Karl Barth, quoting Bengal, said of the entire tradition of Jesus' life is certainly true of the events of the Markan passion: even in their darkest moments, *spirant resurrectionem* (*Dogmatics in Outline*, 102).

BIBLIOGRAPHY

In general on oral characteristics in Mark's narration of individual episodes, see Werner H. Kelber, *The Oral and the Written Gospel: The Hermeneutics of Speaking and Writing in the Synoptic Tradition, Mark, Paul, and Q* (Philadephia: Fortress, 1983). On pronouncement stories and rabbinic disputations, see Philip S. Alexander, "Rabbinic Biography and the Biography of Jesus: A Survey of the Evidence," in C. M. Tuckett, *Synoptic Studies: The Ampleforth Conferences of 1982 and 1983* (Sheffield: JSOT, 1984), 41–42; on pronouncement stories as *chreiai*, see also the bibliography for Chapter 5.

On parables generally, C. H. Dodd, *The Parables of the Kingdom* (London: Nisbet, 1936) and Joachim Jeremias, *The Parables of Jesus* (London: SCM, 1976; 1st German ed., Zurich: Zwingli, 1947) were, of course, seminal; and there were important developments in the subsequent studies of Eta Linneman, *Jesus of the Parables: Introduction and Exposition* (New York: Harper and Row, 1966), Daniel Otto Via, *The Parables: Their Literary and Existential Dimension* (Philadelphia: Fortress, 1967), and John Dominic Crossan, *In Parables: The Challenge of the Historical Jesus* (New York: Harper and Row, 1973). On parables in an oral context, see particularly Barbara Kirshenblatt-Gimblett, "A Parable in Context: A Social Interactional Analysis of Storytelling Performance" in Dan Ben-Amos and Kenneth S. Goldstein, eds., *Folklore Performance and Communication* (The Hague and Paris: Mouton, 1975). For more recent work: on parables in Mark, see John R. Crossan, S.J., *The Gospel in Parable: Metaphor, Narrative and Theology in the Synoptic Gospels* (Philadelphia: Fortress, 1988), especially 28–62, and literature there cited; on parables' demand for responsive action (that is, parables as deliberative rhetoric) see Bruce Chilton and J. I. H. McDonald, *Jesus and the Ethics of the Kingdom* (London: SPCK, 1987) especially 14–43, 63–76.

On the oral style of Markan summary passages, see Joanna Dewey, "Oral Methods of Structuring Narrative in Mark," *Interpretation* 43.1 (1989): 32–44. Dewey disagrees here with Kelber, who argues that individual episodes, rather than Mark as whole, show the signs of oral style (*The Oral and the Written Gospel*, 44–80).

For Werner Kelber's view of Mark's passion and resurrection narratives, see *The Oral and the Written Gospel*, 185–99. Kelber understands Mark's passion narrative as "truly about passion and death, not passion and resurrection" (186); one "cannot possibly infer" from the conclusion to Mark's narrative "a happy resolution for the disciples" (186). I disagree so totally that a reader might be forgiven for wondering if we were speaking of the same text; and it is, of course, finally the reader's experience of the text that must decide. See also bibliography to Chapter 9.

11

As It Is Written: Oral Characteristics of Mark's Appeals to Scripture

Scarcely has Mark gained our attention and begun his narrative, when he tells us that what we are to hear took place "as it is written" (1:2). As I have claimed to be considering the *oral* characteristics of Mark's narrative, that I should now examine his appeal to written authority may seem surprising. Surely such an appeal is essentially alien to orality? In the whole of Homer, we may recall, there is only one possible reference to a written message, and even on that occasion it is not entirely clear that the "baneful signs" (σήματα λυγρά) given by Proteus to Bellepheron as a means of compassing his death actually involve genuine alphabetic writing (*Iliad* 6.168). Against this, however, students of oral traditional literature may point to the example of Caedmon's Hymn, here quoted in full:

> Nu sculon herigean heofonrices weard,
> meotodes meahte and his modgeþanc,
> weorc wuldorfæder, swa he wundra gehwæs,
> ece drihten, or onstealde.
> He ærest sceop eorðan bearnum
> heofon to hrofe, halig scyppend;
> þa middangeard moncynnes weard,

ece drihten, æfter teode
firum foldan, frea ælmihtig.
(West Saxon version, typographically normalized)

[Now we must praise the Ruler of Heaven
The might of the Lord and His purpose of mind,
The work of the Glorious Father; for he,
God eternal, established each wonder,
He, Holy Creator, first fashioned the heavens
As roof for the children of earth.
And then our guardian, the Everlasting Lord,
Adorned this middle-earth for humankind.
Praise the Almighty King of Heaven.]
(trans. Kevin Crossley-Holland, 1965)

Here, clearly, is a poem filled with echoes of scriptural and Christian tradition. Yet this, as Donald Fry has shown us, is one of the few Old English poems that we know with virtual certainty to have been composed orally, and by an illiterate ("Caedmon as a Formulaic Poet" and "The Memory of Caedmon"). Caedmon, of course, had acquired his knowledge by listening to the learned monks with whom he daily associated. What we have in Caedmon is, in fact, a special case of that general principle of transmission in a chirographic society noted earlier: that publication usually means performance, and that most people experience literature by hearing it. What we must do, then, is examine *how* Mark uses Scripture.

Mark's Use of Scripture: 1:1–13

Prologue. Witness to the Coming One: In the Wilderness (1:1–8)

Mark's first words assert that Jesus fulfils Jewish eschatological hope, for the evangelist speaks of him as "Christ" (1:1). We should not, of course, import into this term precisely the implications that it was to have either in later Jewish or later Christian tradition. In Judaism of the first century "anointed" was used variously (compare de Jonge, "Messianic Ideas"). It *is* nevertheless, a biblical term, and it was used of God's agent, or agents, of final deliverance, and that is what concerns Mark here.

More precisely, Mark says that the "beginning of the good news of

Jesus Christ" was the coming of the Baptiser in fulfilment of prophecy: and to prove it he cites (more or less) Exodus 23:20, Malachi 3:1a, and Isaiah 40:3. These passages have, apparently, a history of interpretation in connection with Israel's hope. Combining Exodus 23:20 with Malachi 3:1 seems to have been traditional (compare Matt. 11:10 and Luke 7:27; note also *Exodus Rabbah* on 32.9). Isaiah 40:3 had been applied by the Qumran community to their own withdrawal into the wilderness, which they certainly regarded as a decisive moment in God's dealings with Israel (compare 1QM 1.3; 4QpPs[a] 3.1; 4QpIsa[a] 2.18); the same passage of Scripture is alluded to in an eschatological setting in the opening chapters (probably late pre-Christian) of *1 Enoch*. In the last time, says the writer, "Mountains and high places will fall down and be frightened. And high hills shall be made low" (*1 Enoch* 1.6; compare Isa. 40.4). Yet even though Mark's choice of texts links him with the scribes of Qumran and the later rabbis, his manner of handling the material is quite different from theirs, being similar, in fact, only to the allusion in *1 Enoch*. As we have said, Mark "cites (more or less)"—and that is the most we can say. Even allowing for the multitude of versions available in antiquity, it cannot be claimed that he quotes from any of them: what he offers is not quotation, but reminiscence. Indeed, he does not even trouble to check the source of his allusions, but attributes them all to Isaiah (1:2). In short, his use of the tradition appears (like Caedmon's) to owe more to insight and thoughtful listening than to careful study or academic discussions in a school of interpretation.[1]

It is no surprise that the bearer of God's message appears "in the wilderness" (1:4), for in Jewish tradition that is naturally the place of refuge and hope. Moses, David, and Elijah all fled to the desert (Exod. 2:15; 1 Sam. 23:14; 1 Kgs. 19:3–4; compare Ps. 55:6–8; 1 Macc. 2:27–28; 2 Macc. 5:27; 1 QS 8.13 [compare 9.19–20]; *PsSol.* 17.16–17; Josephus, *Life* 11–12). The Baptist is, Mark tells us, austerely dressed;

1. Augustine Stock attempts to explain the wrong citation at 1:2 as an example of Markan "bracketing" (see *Call to Discipleship*, 33). This cannot, I think, be accepted. Bracketing [*inclusio*] means surrounding one literary element with another for a particular effect; it does not mean saying that you are going to do one thing, and then doing something else. In any case, this is not Mark's only erroneous reference to Scripture. At 2:26 he has Jesus claim that David ate the bread of the presence "when Abiathar was high priest"; according to 1 Samuel 21:1–6, 22:20, the high priest who helped David was Ahimelech, who was the father of Abiathar.

indeed, the description is quite elaborate and initially surprising in a writer who does not trouble to tell us what Jesus looked like. It will have reminded those sensitive to the language of biblical tradition that the Baptist is a man of God and a prophet (compare Zech. 13:4; Dan. 1:12–17; Josephus, *Life* 11, 14; Philo, *Vita Cont.* 37; *TIsaac* 4.5; *1 Clem.* 17:1). Considered aurally, however, it was even more precise in its associations than that. Greek-speaking Jews who had listened to the Septuagint in synagogue would have heard it speak of Elijah who wore not only a prophet's hairy mantle but also a leather girdle, and might well catch a verbal echo in Mark's description of the Baptist:

4 Kingdoms 1:8 LXX	Mark 1:6
and a leather girdle	and a leather girdle
girded around	around
his waist	his waist
καὶ ζώνην δερματίνην	καὶ ζώνην δερματίνην
περιεζωσμένος	περί
τὴν ὀσφὺν αὐτοῦ	τὴν ὀσφὺν αὐτοῦ

When King Azariah heard those words he said, "It is Elijah the Tishbite" (2 Kgs. 1:8): and so, perhaps, did some of Mark's auditors (compare 9:11–13). Some circles in Judaism, at least, had long known that the prophet Elijah's return would precede the final "day" of God (Mal. 4:5; Sirach 48:10); and there is evidence of rabbinic teaching going back to the time of Herod, according to which Elijah would come to put right injustices perpetrated by violence (*m.* 'Eduyyot 7:1)[2]. It is in the light of such hopes as these that Mark's description of the Baptist is to be understood. The evangelist is not primarily concerned with telling us about the Baptist at all: he is telling us again that in the coming of Jesus, God's "day" has dawned. Again, however, the allusion is purely a matter of oral reminiscence: the older text is not even specifically mentioned.

The Baptism of Jesus (1:9–11)

Mark first tells us of Jesus' baptism in plain language, "In those days Jesus came from Nazareth of Galilee and was baptized by John in the

2. Cited by Ephraim E. Urbach, *The Sages: Their Concepts and Beliefs* (Jerusalem: Magnes, 1979), 660–61; compare 995, note 45.

Jordan" (1:9). Then he sets it in the world of apocalyptic revelation: "And when he came up out of the water, immediately he saw the heavens opened, and the Spirit descending upon him like a dove" (1:10). The outpouring of God's spirit on the ideal king is an idea at least as old as Isaiah (compare Isaiah 11:2), and appears as a later expression of eschatological hope in Judah's words in the *Testaments of the Twelve Patriarchs* (late second or early first century B.C.E.): "And the heavens will be opened upon him to pour out the spirit as a blessing of the Holy Father" (*T12P TJud* 24.2; compare *T12P TLevi* 18:6).

But there was, Mark tells us, even more: "a voice came from heaven, 'Thou art my beloved Son; with thee I am well pleased'" (1:11). Heavenly voices in support of righteous persons or in solution to problems are not uncommon in the Hebrew Scriptures and other early Jewish writing (for example, Gen. 21:17, 22:11, 15; Dan. 4:31; 1 *Enoch* 65.4). They also occur in later rabbinic literature (for example, *b. Ta'an* 24b). The latter, however, tends to a characteristically academic caution about such manifestations. They are well enough in their place, but they are not to be compared to the work of scholars (*t. Sota* 48b; *b. Yoma* 9b)! Mark, by contrast, clearly regards the voice as of unquestioned authority (compare 9:7).

It became evident in our study of Markan structure that the notion of Jesus as God's son is important for Mark, and is emphasised at key moments in the narrative ([1:1,] 1:11, 9:7, 12:6, 14:61–2, 15:39). Obviously, to speak of Jesus in this way is to imply that he is close to God (compare *JosAsen* 18:11, 21:4). For the ear of a Hellenistic Jew who was used to listening to the Septuagint, however, when the story of Jesus' baptism referred to him as God's son there was yet another striking aural reminiscence—this time from the story of Abraham and Isaac in Genesis 22. The Greek text of Genesis twice paraphrases God's words to Abraham (Hebrew: בִּנְךָ אֶת־יְחִידְךָ RSV, "your only son") by τὸν υἱόν σου τὸν ἀγαπητόν,(your beloved son) (LXX Gen. 22:2, 16), which is exactly echoed by Mark's ὁ υἱός μου ὁ ἀγαπητός(my beloved son) (1:11).

In Jewish tradition, there were certainly parallels to the notion of describing God's relationship to "the anointed" in terms of Abraham's relationship to Isaac. *Testaments of the Twelve Patriarchs*, for example, has Levi speak of the coming "new priest, to whom all the words of the Lord will be revealed" whose "star shall rise in heaven like a king:"

> The heavens will be opened,
> and from the temple of glory sanctification will
> come upon him,
> with a fatherly voice, as from Abraham to Isaac.
> And the glory of the Most High shall burst forth upon him.
> And the spirit of understanding and sanctification
> shall rest upon him. (*T12P TLevi* 18.6)[3]

Mark, however, has chosen specifically to echo that part of the Genesis narrative which tells of the offering or "Binding" (Hebrew, *Aqedah*) of Isaac. Recent studies of the *Aqedah* have gone far in estimating its significance for early Judaism, elevating it almost to the status of a Jewish doctrine of the Atonement. This is to place much weight upon relatively few texts; but that the Binding of Isaac was occasionally understood as an offering for the sake of Israel cannot be denied. It is spelled out, for example, in the *Antiquities of Pseudo-Philo*, a nonschool writing likely to reflect the milieu of the Palestinian synagogues at the beginning of the first Christian century:

> [God said]: "I demanded [Abraham's] son as a holocaust. And he brought him to be placed on the altar, but I gave him back to his father and, because he did not refuse, his offering was acceptable before me, and on account of his blood I chose them." (*Ps. Philo* 18.5; compare 32.1–4; see also 40.2, which [in line with a somewhat feminist tendency of *Pseudo-Philo*] explicitly compares the sacrifice of Jephthah's daughter with the Agedah)

Perhaps such ideas lie behind Mark's description here. Does he mean us to understand that Jesus was, as the (possibly late first century) *Epistle of Barnabas* puts it, "a sacrifice, in order that the type established in Isaac, who was offered upon the altar, might be fulfilled?" (7.3). Does Mark also see Jesus as God's Isaac—the lamb provided by God? Again, it is all a matter of echo and allusion: there is no question of explicit citation or exegesis.

The Testing of Jesus (1:12–13)

After Jesus' baptism, Mark (1:12) tells us, "The Spirit immediately drove him out into the wilderness." The wilderness is a place of ref-

3. The passage may contain an allusion to one of the Maccabean priest kings: see H. C. Kee in J. H. Charlesworth, *Old Testament Pseudepigrapha*, vol. 1, 794, notes 18a,b. Compare also John 1:32–33.

uge and hope: it is also dangerous, outside the boundaries and restraints of ordinary society, the abode of demons (Lev. 16:10; Tob. 8:3; *1 Enoch* 10.4–5; *2 Bar.* 10.8). As God's son Israel was sent into the wilderness and tested forty years after the Exodus, so God's son, Jesus, is driven into the wilderness forty days to be tested. Jesus was, says Mark, "tempted by Satan; and he was with the beasts, and the angels ministered to him" (1:13). The connection has been disputed, and certainly Mark does not emphasize it, but personally I find it hard to see how anyone familiar with interpretative traditions surrounding Genesis 1–3 such as those reflected in the *Life of Adam and Eve* and the *Apocalypse of Moses*[4] could avoid hearing echoes of them in this. In three notable ways such traditions offer parallels to Mark's narrative. First, they identify Satan himself, humanity's determined adversary, with the tempter (*LAE* 9.1; *ApocMos.* 17:1–5; compare Rev. 12:9). Second, just as Jesus is "with the beasts," so these traditions emphasize the relationship of unfallen humanity to the animals—a relationship so close that in some measure it even survives Adam's disobedience. Thus, after the expulsion from Paradise, "all the angels and all the creatures of God surrounded Adam as a wall around him, weeping and praying to God on behalf of Adam, so that God gave ear to them" (*ApocMos.* 29.13–14; compare *LAE* 8.1–3). Third, such traditions speak of the closeness of humanity's relationship to the angels. Before the disobedience, Adam and Eve fed on the food of angels (*LAE* 4.2; compare *b. Sanh.* 59b). Notably, the word used by Mark to speak of the angels' "ministry" to Jesus is διακονέω (transliterated *diakoneō*), a verb sometimes associated with service at table on formal and ceremonial occasions (for example, 1:31; Acts 6:2).[5] Perhaps then for Mark as for Paul (Rom. 5:15–21; 1 Cor. 15:21–22, 45–49) Jesus is a second Adam,

4. Despite the way in which the *Life of Adam and Eve* and the *Apocalypse of Moses* are frequently presented and discussed (for example, H. F. D. Sparks, ed., *The Old Testament Apocrypha* [Oxford: Clarendon, 1984], 141–67; James H. Charlesworth, ed., *The Old Testament Pseudepigrapha*, vol. 2 [New York: Doubleday, 1985], 249–95), they are in fact quite distinct documents (compare John R. Levison, *Portraits of Adam in Early Judaism: From Sirach to 2 Baruch* [Sheffield: JSOT, 1988], 163–90), although they do contain some of the same traditions.

5. See John N. Collins, *Diakonia: Re-interpreting the Ancient Sources* (New York and Oxford: Oxford University Press, 1990), 150–68.

regaining for humanity the honors lost by the first? Again, it is all a matter of allusion and reminiscence.

Implications of 1:1–13

Our examination of Mark's opening narratives suggests that we are dealing with a work deeply imbued with scriptural tradition and traditions of scriptural interpretation, yet devoid of academic or literary pretension. Mark does not quote texts from Hebrew Scripture as one who has studied them, he alludes to them as one who has heard them quoted and remembered them. It is notable that two of the most striking scriptural reminiscences considered above (the description of the Baptiser, and Jesus as God's "beloved Son") depend entirely on aural reminiscence for their effect.

Allusion and Reminiscence

This allusive way of using Scripture as it has been heard and remembered is characteristic of Mark, and can easily be illustrated from other key features and sections in the text.

The Kingdom of God

Mark begins Jesus' public ministry with his proclamation of the imminent kingdom of God (1:15). As we have seen, according to Mark, those who are disciples of Jesus have already "been given the secret of the kingdom of God." If, moreover, they will confess the crucified and risen Christ, and if they will be willing to suffer with him, they will see the kingdom of God "in power" in their own lives (4:11, 9:1). But where does the idea of "the kingdom of God" come from? Mark seems to assume that his audience will understand it, and never bothers to explain it.

On the basis of a careful examination of parts of the Isaiah Targum, Bruce Chilton (*Glory of Israel*) has argued that Markan usage (and synoptic usage generally) was similar to that of the first-century Palestinian synagogue. Using the phrase "kingdom of God" to speak of God's active, personal intervention to redeem and restore Israel was characteristic of ways in which the synagogue understood texts such as Isaiah 24:23, where the Hebrew has,

> the LORD of hosts reigns on Mount Sion,

and the Targum says,

> the kingdom of the LORD of hosts will be revealed on Mount Zion.

Or Isaiah 31:4, where the Hebrew has,

> the LORD of hosts will descend to fight upon Mount Zion,

and the Targum says,

> the kingdom of the LORD of hosts will be revealed to dwell upon Mount Zion.

If Chilton is correct, Mark's use of the phrase "kingdom of God" implies and is based upon scriptural exegesis. The association of the kingdom of God with the person of Jesus Christ means that in and through the crucified and risen One we are met by God's active, personal intervention to redeem and restore Israel (and the world). Yet never once does Mark quote a text of Scripture to illustrate this, or show Jesus doing so.

Son of Man

At various points in Mark's narrative, Jesus speaks of himself as "Son of man" (2:10, 2:28, 8:31, 8:38, 9:31, 10:33, 10:45, 13:26, 14:62). No one else in the narrative ever speaks of him in this way. Problems surrounding the history and original meaning of the phrase are enormous, and it will certainly not be appropriate to go into them here. As far as Mark's own understanding is concerned, however, Morna D. Hooker's analysis in *The Son of Man in Mark* (1967) still appears to be essentially correct. What links all "Son of man" sayings in Mark is "the question of Jesus' authority—the authority which he claims and which his disciples accept" (179).

Thus, it is while claiming authority to forgive sins, and while claiming Davidic authority as lord of the Sabbath, that Jesus first calls himself "Son of man" in controversy with his opponents (2:10, 27–28). It is in connection with the disciples' acceptance of his authority as God's messiah that he later uses it to speak of his coming suffering, and also to speak of their share in that suffering if they choose to follow him (8:31, 9:31, 10:33, 10:45). It is in connection with suffering—his own and his disciples'—that he uses it to speak of his final authority in judgment (8:38, 13:26). It is in this connection, finally, that he again calls himself "Son of man" publicly at his trial, when he replies to the high priest's question (14:62).

"Son of man" in Mark speaks, then, of Jesus' authority. Two things more may be said. First, that for Mark the title seems to speak particularly of Jesus' Davidic, messianic authority. Once Jesus is shown using this authority specifically in connection with a claim to act as David did (2:27); twice he is shown using it in direct connection with the messianic claim (8:31, 14:62). Second, we should note that "Son of man" seems in Mark's view to have a connection with Daniel 7:14, since Jesus is shown three times using the title in obvious allusion to this text, and always at important moments: at Caesarea Philippi (8:38), in the farewell discourse (13:26), and in his response to the high priest at the trial (14:62). What is the force of this allusion? The matter is debated, but the following considerations appear to be relevant.

Examples of allusion to Daniel 7 in Jewish material from our period survive in the "Book of the Similitudes" (*1 Enoch* 37–71), 2 (4) Esdras (both probably to be dated in the late first century C.E.), and perhaps in *2 Baruch* (probably early second century). In *1 Enoch* 71, Enoch himself is heard describing his own translation into heaven:

> Then the Ancient of Days came with Michael, Gabriel, Raphael, Phanuel, and a hundred thousand and ten million times a hundred thousand angels that are countless. Then an angel came to me and greeted me and said to me, "You, Son of man, who are born in righteousness and upon whom righteousness has dwelt, the righteousness of the Ancient of Days will not forsake you. . . . For from here proceeds peace since the creation of the world, and so it shall be unto you for ever and ever. Everyone that will come to exist and walk shall follow your path, since righteousness never forsakes you." (*1 Enoch* 13–16)

Enoch himself, apparently, is here seen as the "Son of man" who has been exalted to heaven and who will be final judge of the world (compare Gen. 5:24).

Apparently influenced by the imagery of Daniel 7:14 is the reference in *2 Baruch* to a time after completing his work, "when the time of the appearance of the Anointed One has been fulfilled, and he returns in glory"—presumably to heaven whence he came (*2 Baruch* 30:1). Rather more clearly, in 2 (4) Esdras 14:9, the seer is told that he will be "taken up from among humankind, and henceforth you shall live with my Servant (that is, the messiah)."

Taken together, these three sources suggest that use of the "Son of man" image from Daniel 7:14 to speak of an (or the) eschatological savior was not unknown in Jewish tradition of Mark's period. Consistently with this, later rabbinic usage also regularly associates Daniel 7 with the king-messiah. By way of example, we cite a passage ascribing messianic interpretation of Daniel 7:9 to Rabbi Akiba. Of the "thrones" in Daniel 7:9, Akiba was said to have affirmed,

> "one for him [God], and one for David." Rabbi Yosi the Galilean said to him, "Akiba! How long will you treat the divine presence as profane? Rather [it must mean] one for justice, and one for grace" (*b. Chagigah* 14a).

Rabbi Yosi objects to the dangerous implications of David occupying a second throne beside God, but he does not question the messianic interpretation of the "one like a son of man."

So when Mark understands the "Son of man" imagery of Daniel 7 as referring to messianic authority, he appears to be in general accord with a tradition of scriptural interpretation. Yet again (like *1 Enoch* and the other apocalyptic writers, and in contrast to the rabbis) Mark never explicitly names or quotes his biblical source. His references are, as usual, all matters of allusion and reminiscence.

The Passion

Mark's passion narrative is replete with references to the Scriptures. Howard Clark Kee, in a detailed and careful analysis, finds in chapters 11–16 no less than 57 quotations, about 160 allusions, and 60 examples of influence ("Function of Scriptural Quotations," 167–71). And yet, just as was the case with Mark 1:1–14, anyone who examines Mark's text of the Passion with a Septuagint or Hebrew Bible in hand will find that there is (even by the somewhat elastic standards of the first century) scarcely ever an exact quotation, and never a lengthy one in the manner of Matthew or the Letter to the Hebrews. Even key allusions such as those at the Last Supper, to Exodus and Jeremiah (14:24; compare Exod. 24:6–8; Jer. 31:31–34), or at the Trial, to Daniel and Psalm

110 (14:62; compare Dan. 7:13; Ps. 110:1) are all indirect and imprecise: we are dealing, yet again, with matters of memory, not textual comparison.

The Resurrection

Mark's final episode at the empty tomb (16:1–8) is extraordinarily restrained and severe, as is shown even by comparison with its parallels in the other gospels (Matt. 28:1–8; Luke 24:1–9; John 20:1, 11–13). Yet for all its brevity and simplicity, Mark has given us an angelic vision soaked in biblical reminiscence: a thoroughly Jewish way to describe a revelation from God. Among its striking features we note the description of the heavenly messenger as a "man"; his appearance, position, and clothing (the "right hand" designating authority); the initial awe of those who receive the vision; the reassurance offered by the messenger; the announcement of God's gracious and redeeming act;[6] the promise of a sign; and the continuing awe of the recipients. These features, in various combinations, are all regularly parallelled in accounts of angelic visions and visitations in the Old Testament (compare Judg. 6:11–24, 13:3–23; Dan. 10:1–21; 2 Macc. 3:24–34; Tobit 5:1–12, 12:6–20) and in related literature (compare *JosAs* 14:11–15:12; Luke 1:8–20, 26–38). It is in the light of this scriptural tradition that Mark's hearers are able to understand the significance of the vision and to know that the message of the cross was true: the crucified Jesus has been vindicated as he said he would be: "you seek Jesus of Nazareth who was crucified . . . he is going before you to Galilee; there you will see him, as he told you" (16:7).[7] Once again, however, it is all done by allusion and echo. The episode contains not one direct scriptural quotation from beginning to end.

6. In contrast to RSV ("He has risen"), NRSV is surely right to translate ἠγέρθη (16:6) as an aorist passive—the passive voice reverently veiling the divine activity. "It is," as Morna Hooker observes, "God who has raised Jesus to life, and it is his mighty act that is announced to the women" (*Gospel According to St. Mark*, p. 385).

7. Notable is the insistence in the (late/first or early second century C.E.) *Kerygma Petrou* that Jesus' coming, passion, and exaltation were found "in the books of the prophets . . . how all was written that he had to suffer" and that "we recognize that God enjoined them, and we say nothing apart from Scripture" (Dobschütz, 1893, fragments 9 and 10; Schneemelcher, 1965, 102). I am grateful to Ellen Aitken for drawing my attention to this characteristic of *Kerygma Petrou* in her unpublished paper, "The Morphology of the Passion Narrative: How Did Early Christians Tell a Story of Jesus' Sufferings?"

Mark's Two Precise Quotations

Citing example after example to illustrate the same point grows tedious. Anyone who reads Mark through, comparing it either with the Septuagint or a Hebrew Bible, will see that, for all its frequent scriptural allusions, it contains almost no exact or extended quotation. The most that is achieved by way of accurate quotation is, twice, a few lines. These two mild exceptions to the general rule are worth examining. First, in the account of Jesus' dispute with the Pharisees over *Qorban* (7:1–13), we have (following the arrangement in Nestle-Aland 25) four lines quoted from Isaiah 29:13:

> This people honors me with their lips,
> but their heart is far from me;
> in vain do they worship me,
> teaching human precepts as doctrines.
>
> Οὗτος ὁ λαὸς τοῖς χείλεσίν με τιμᾷ,
> ἡ δὲ καρδία αὐτῶν πόρρω ἀπέχει ἀπ' ἐμοῦ·
> μάτην δὲ σέβονταί με,
> διδάσκοντες διδασκαλίας ἐντάλματα ἀνθρώπων.
> (7:6b–7; compare Isa. 29:13 LXX).

Of these four lines, although the first and fourth are free paraphrases that match neither the LXX nor the Masoretic text nor the Targum, the second and third do follow the LXX exactly. Even in this case, however, we note that the text being cited involved parallelism and was therefore easy to remember. Moreover, this particular passage seems likely to have had some currency, and hence increased familiarity, as a Christian proof text: compare Colossians 2:22, *1 Clement* 15.2, and *The Shepherd of Hermas*, Mandate 12.4.4.

Second, at the end of the Parable of the Wicked Husbandman, Mark gives us four lines quoted exactly from the LXX:

> The very stone which the builders rejected
> has become the head of the corner;
> this was the Lord's doing,
> and it is marvellous in our eyes.
>
> Λίθον ὃν ἀπεδοκίμασαν οἱ οἰκοδομοῦντες,
> οὗτος ἐγενήθη εἰς κεφαλὴν γωνίας·
> παρὰ κυρίου ἐγένετο αὕτη,

καὶ ἔστιν θαυμαστὴ ἐν ὀφθαλμοῖς ἡμῶν.
(12:10–11; compare Ps. 117:22–23 LXX [MT 118:22–23]).

This is by far Mark's most impressive example of reasonably accurate biblical quotation. Again, however, we note that the text involves parallelism and was therefore easy to remember, and that it, too, seems likely to have had currency and therefore familiarity as a Christian proof text: see Acts 4:11 and 1 Peter 2:7 (both passages, interestingly enough, connecting the allusion with Peter), the *Epistle of Barnabas* 6.4 (which also understands it as speaking of Christ's passion) as well as the *Gospel of Thomas* 66, which, of course, associates it with the same parable as does Mark.

Scriptural Tradition as a Means of Articulation

It is evident from Mark's text that its author is familiar with Scripture and with certain traditions of interpreting Scripture. On the other hand, the text of Mark offers us no certain evidence that its author had ever actually *read* any Scripture. I do not deny this possibility; I simply point out that on the basis of Mark's text, it could not be established. Every word of scriptural tradition and allusion in Mark could have been acquired through oral transmission. Even Mark's mistakes (1:2–3, 2:26) are of a type that could perfectly well come about through mishearing, rather than misreading.

We have only to compare Mark's use of the Scriptures with Matthew's, or the writer to Hebrews', or even Luke's, to see what happens when a more literate or scribal mind takes hold of the traditions. Matthew's opening chapter, with its careful numerology, can only have been worked out as a written text with other written texts to hand. It would be inconceivable in Mark. Matthew and Luke, moreover, will both correct Mark's careless citations (1:2–3, compare Matt. 3:3, Luke 3:4–5; for 2:26, compare Matt. 12:1–8, Luke 6:1–5). It remains, nonetheless, that in his own way and after his own fashion Mark appeals to Scripture constantly. Why? For one who has heard Scripture Sabbath by Sabbath, Scripture's language is the natural language of divine event. Therefore, any event viewed as significant, whether it is the fall of Jerusalem or the passion and resurrection of the Lord, is inevitably and spontaneously spoken of in that language.

A simple but highly illustrative parallel to Mark's use of Scripture is provided by the Homer scholar Milman Parry's account of an interview in 1933 with Mitcho Savitch, a Yugoslavian oral (nonliterate) bard who was then eighty-two years old. Savitch told Parry the story of his life, beginning with an event of which he was obviously extremely proud, namely, the part he had personally played in the Battle of Ravno. Parry tells of listening to the old man's account. What Parry found striking was that it repeatedly fell into the forms of heroic narrative verse.

> Dvije paše bismo i ubismo,
> A Selima živa ufatismo.
> *Two pashas we fought and overcame,*
> *And Selim we took alive. . . .*
> It must have been a very old verse . . . it is still the form in which an old man casts the thought of his own life. It is no verse that he has made, but has come down to him from the past. For the people as a whole who created the verse and kept it, it is an ideal; for this man it has become a boast. And as we can see in the case of this one verse, so the whole body of traditional poetry from the past brought with it the ideal of life as a whole for these men of Gatsko who have ever been renowned for their singing. So in the Greek heroic age did they sing the κλέα ἀνδρῶν—*the high deeds of men.* (Parry, *Homeric Verse*, 390)

Just so, those who inherited a tradition greater than that of Gatsko or even Homer found it natural and necessary to speak in terms of that tradition, if they were to express properly what was most important and most true for them. That is why Mark can never let us forget that the story he tells took place "as it is written."

BIBLIOGRAPHY

On Caedmon's Hymn, see Donald K. Fry, "Caedmon as a Formulaic Poet," in *Forum for Modern Language Studies* 10 (1974); reprinted in Joseph J. Duggan, ed., *Oral Literature: Seven Essays* (Edinburgh: Scottish Academic Press; New York: Barnes and Noble, 1975), 41–46. See also Donald K. Fry, "The Memory of Caedmon," in John Miles Foley, ed., *Oral Traditional Literature: A Festschrift for Albert Bates Lord* (Columbus, Ohio: Slavica, rpt. 1983) 282–93.

On Mark's use of Scripture in connection with orality, there are useful comments in Robert M. Fowler, *Let the Reader Understand: Reader-Response Criticism and the Gospel of Mark* (Minneapolis, Minn.: Fortress, 1991), 88–90.

On "the Christ," see Marinus de Jonge, "Messianic Ideas in Later Judaism," Parts 1,2,3 and 5 in Gerhardt Friedrich, ed., *Theological Dictionary of the New Testament*, vol. 9 (Grand Rapids, Mich.: Eerdmans, 1974), 509–17, 520–21; and Adam Simon van der Woude, "Messianic Ideas," Part 4 in Friedrich, op. cit., 517–20.

On the Aqedah, see Geza Vermes, *Scripture and Tradition in Judaism* (Leiden: Brill, 1973), 193–227; for criticism of Vermes's method and some of his conclusions, see E. P. Sanders, *Paul and Palestinian Judaism: A Comparison of Patterns of Religion* (London: SCM, 1977), 28–29.

On "kingdom of God" in the first-century Palestinian synagogue, see Bruce Chilton, *The Glory of Israel* (Sheffield: JSOT, 1983).

On Markan use of the phrase "Son of man," see Morna D. Hooker, *The Son of Man in Mark: A Study of the Background of the Term "Son of Man" and Its Use in St. Mark's Gospel* (London: SPCK, 1967), 174–82; contrast, however, Douglas R. A. Hare, *The Son of Man Tradition* (Minneapolis: Fortress, 1990), 183–211.

On Mark's use of scriptural allusions in his Passion narrative, see Howard Clark Kee, "The Function of Scriptural Quotations and Allusions in Mark 11–16," in E. Earle Ellis and Erich Grässer, eds., *Jesus und Paulus: Festschrift für Werner Georg Kümmel*. Göttingen: Vandenhoeck and Ruprecht, 1975.

12

Conclusions: Mark in Its Setting

A Writer Who Wrote to Be Heard

Was Mark written to be read aloud? Everything suggests that the answer to that question must be yes. Mark was designed for oral transmission—and for transmission as a continuous whole—rather than for private study or silent reading. When we examined Mark's structural arrangements, again and again we found ourselves dealing with broad thematic effects that would emerge naturally in the course of a performance of the whole, but that can hardly emerge otherwise. Subsequent examination of the details of Markan style, and even his use of Scripture, all reinforced this impression. Nonetheless, the proof of this particular pudding must be in the eating. Will Mark work as we say it would? According to a number of critics, that was exactly how Mark *did* work, when performed by a competent professional actor who knew how to interpret and use the opportunities it gives. This is how Janet Karsten Larson described a performance by the British actor Alec McCowen:

> McCowen's deliberate repetition of significant gestures, of facial changes, of shifts in tone, and of certain blocking patterns brings vividly before our attention Mark's thematic design. . . . Jesus' prescience about his fate, the disciples' lack of comprehension, their master's corrective teaching about discipleship, . . . In McCowen's enactment

> these rhythms are unmistakably felt. . . . Seeing it all done so compellingly, one marvels that no one thought to play this Gospel straight before. For Mark's work *is* dramatic; McCowen simply calls our attention to what we should have guessed. ("St. Alec's Gospel," 17)

Was Mark's Gospel Composed Orally?

Is the effect of the features we have observed such as to lead us to go further? Should we regard the gospel as the textualization of a "life" which was also composed orally? Are we dealing, not merely with oral transmission, but with oral traditional literature in the full sense? This suggestion has been put forward, not just for Mark but for the gospels generally, by no less an authority than Albert Bates Lord himself ("The Gospels as Oral-Traditional Literature"), and therefore must be taken seriously. It remains unlikely, I think, for three reasons.

First, as Lord concedes, "in their normal tellings, oral traditional narratives about individuals, whether in verse or prose, only rarely include a single account that begins with birth and ends with death. Most commonly, the separate elements or incidents in the life of the hero form individual poems or sagas" (39–40). The Hellenistic "life" is a product of literacy, though still deeply indebted to oral habits of mind. Second, granted that oral memory could sustain narratives as long as Mark (and it certainly could), and granted that Mark does provide the kind of narrative favored by orality; still, oral memory seems to show this capability under certain rather clearly identifiable conditions of preliteracy that Lord and others have themselves described. As we have observed, there is no reason to suppose that the gospel traditions arose or were handed on in anything approaching a preliterate culture; rather the reverse. It is not without reason that Hellenism has been called the civilization of *paideia.* The gospels could hardly have been composed at a period of antiquity when there was more general literacy and education. Third, the sustained verbatim textual similarities between the three synoptic gospels do not appear to be characteristic of oral traditional repetition of the same material, and seem to many (including myself) to require that we posit some kind of literary utilization theory to explain them. As Lord admits, "on occasion the texts are so close that one should not rule out manuscript transmission" (90).

As students of oral traditional literature are increasingly emphasizing, we may never assume that a piece of literature was composed orally, just because it has oral features. We need always take into account what we know of the circumstances of its production, too. Old English poetry such as *Beowulf* also has many oral features, notably what appear to be oral formulaic building blocks. Yet we find exactly similar building blocks in other old English poetry that was evidently composed in writing, some of which was even translated from Latin originals. The same factor has been observed by students of oral tradition in pre-Islamic Arabic poetry, and in Ferdowsi's *Shâhnâma*, the authoritative version of the national epic of Iran. In all cases, whether the poetry in question was written or not, its techniques appear to be oral. Obviously there are differences between these situations and that which we are envisaging for Mark: most strikingly, that in the the former we are dealing with heroic poetry, in the latter with a prose "Life." Yet the essential similarity remains: that in all cases we have writers using oral techniques in composing written texts, partly because their education has been such that these techniques come naturally to them, and partly for the obvious reason that they expect to be heard.

The Author of a Hellenistic "Life"

All witnesses to the spread of the gospel suggest that it involved talk. "Faith," Paul observed "comes from what is heard" (Rom. 10:17). Acts, as we noticed, seems to find quite normal the picture of the apostle talking with his fellow Christians throughout an entire night (20:7–11), and presumably reflects practices that the late first-century author regarded as likely. Clement of Rome (fl. c. 96), describing the work of Jesus' disciples, speaks of their going forth "with glad tidings," of their "preaching," and of their appointing persons to "succeed to their ministry"; nowhere, by contrast, does he mention an obligation to create written witness to Christian tradition, though he was obviously familiar with Christian writings, and valued them (*1 Clement* 42:3, 44:2; compare Ignatius of Antioch, *Philadelphians* 7.2).

On the other hand, in both Jewish and gentile milieus of the period, all this talk would normally have been supported by a degree of notetaking, the notetakers making use not of scrolls, but of parchment notebooks (Latin, *membranae*; Greek, *membranai*: compare 2 Tim. 4:13). Some no doubt made collections of material that seemed naturally to go together, or to bear with particular significance on their understanding of Jesus. Certain kinds of material would inevitably have tended to cluster: collections of sayings, collections of parables, accounts of Jesus in controversy with opponents, accounts of his miracles, and so forth.

Among the early Christian notetakers there was apparently a Greek-speaking Christian of (probably) Jewish descent. He had not received education much beyond the primary stage under a *grammatistēs* (elementary teacher), but he was intelligent, with a good ear and some natural skill at telling a story in simple, popular style. From time to time he had heard various "lives" and accounts of the "Deaths of the Famous" read for edification and entertainment. It was he who determined to make of his own notes and knowledge of the ongoing tradition, and perhaps of earlier collections made by others, a "life" of Jesus that might similarly be read aloud for interest and edification. So much we may reasonably infer. That such inference coincides more or less with Papias's account of Mark as cited by Eusebius need hardly surprise us, though of course it cannot substantiate that account in detail (Eusebius, *Ecclesiastical History* 3.39.15).

If we are correct in our description of the gospel's origins, then Mark's written text would have been understood from the beginning as a recollection of what had been said: it was *hupomnēmata* (memoranda) (Eusebius, *Ecclesiastical History* 2.15.1), *apomnēmoneumata* (recollections) (compare Justin 1 *Apology* 66.3, 67.3–4, *Dialogue* 99–107 [thirteen times]; Eusebius, op. cit., 3.39. 16). Its remembrance of the living voice was decisive for its trustworthiness; that it was a preparation for utterance was the criterion of its usefulness.

A "Scribal" Mark?

To speak further of Mark's setting is, inevitably, to enter the realm of the highly speculative; yet it may be worthwhile briefly to refer to some suggestions that have been made. For all my general admiration of Mary Beavis's careful work, I doubt her view that we should see the evangelist as a scribe or as scribally educated (*Mark's Audience*, 166–70). To begin with, I think Beavis gives Mark rather too high a score for his literary style (42–44), but that alone would not be fatal to her suggestion. More important, as I indicated earlier, there is virtually nothing that is scholastic about Mark's use of Scripture or Jewish tradition. His occasional explanations of Aramaic words (5:41, 7:11, 34, 15:22, 34) and other observations in passing (7:26, 12:42) do suggest some knowledge of Jewish and Palestinian lore, yet involve nothing that any ordinarily intelligent Jew of Palestinian background (such, indeed, as John Mark is described as being in Acts), or even an inquiring gentile, might not have known. By contrast, his erroneous attribution of Pharisaic ritual washings to "all the Jews" (17:3) exhibits the same kind of inaccuracy as his mistaken scriptural attributions at 1:2–3 and 2:26, and his qualification of "the first day of Unleavened Bread" as "when they sacrificed the Passover lamb" (14:12), even if not quite without parallel (see Taylor, *Gospel According to St. Mark*, 535–536) is certainly unusual. Of course we should not impose twentieth-century standards of accuracy on a first-century writer, yet taken together, I find these elements hard to square with scribal training in any formal sense.

Mark the Prophetic Charismatic?

I am also doubtful about Howard Clark Kee's view that the Markan community had an apocalytic, esoteric outlook similar to that of Daniel and Qumran (*Community of the New Age*, 79–87), and that the portrayal of the disciples in the gospel reflects the nature of the Markan community: that is to say, that the Markan community was a band of prophetic, charismatic preachers who repudiated family ties and social involvement, and preferred the countryside to cities, like Jesus and the disciples in the gospel (87–97).

Certainly Mark is deeply concerned, like all early Christians, with eschatology—with the fulfillment of Scripture, the nearness of the kingdom of God, and the coming of the Son of man—and certainly Mark makes occasional use of apocalyptic imagery, notably in his account of the baptism of Jesus, and in the Farewell Discourse. Yet there is nothing particularly esoteric in his presentation of any of these matters. The language of Chapter 13 is obscure to most of us, because we are far removed from the events to which Mark was referring, but there is no reason to suppose that it would have been obscure to Mark's contemporaries. As for the gospel's earlier talk of the "mystery" of the kingdom and the repeated exhortations to silence about Jesus: these are, as we have seen, oral narrative devices to restrain our natural tendency to understand Jesus' words and works without the cross.

As regards the notion that Mark or his community were a group that repudiated family ties: certainly Mark finds the disciples' personal renunciations both admirable and interesting, and sets them forth for our admiration (6:8–11). Yet it is noticeable that those passages in the gospel that speak of the mission and calling of the Twelve are *not* among those to which we referred earlier, where the style of the writing suggests naturally that the reader will invite the audience to feel themselves involved in the advice given. The ancients wrote "lives" in order that their heroes should provide us with moral examples, but they were quite as capable as we are of distinguishing between the essence of a character or a situation, and its particulars. When Aristotle suggested that in seeing a Sophoclean tragedy we experience "pity and fear," "pity" involving an awareness that "this could happen to me," of course he did not imagine that many of us were likely literally to be faced with the problem of having unwittingly killed our fathers or married our mothers (*Poetics* 6.1449^{b}27–28; compare *Rhetoric* 1385^{b}13–20). Clearly he expects us to understand that it is principles that are involved here—the transitoriness of human prosperity, the frailty of human life, and so on—rather than particulars. In exactly the same way, in Mark, a passage such as 8:34–38, where "the multitude" are invited "with the disciples" to take up their cross and follow Jesus, would obviously have been taken as stating a principle rather than as a precept to be followed *au pied de la lettre*. As it happens, some of those in Mark who appear, even if only briefly, as models of discipleship are by no means lacking in family ties (the Syro-Phoenician woman, the father of the epileptic boy); the twelve disciples, by contrast, are almost as often examples of what we should *not* do as of what we *should* do.

As is made clear in the story of the rich young man, Mark is aware of the danger of those riches that make it "hard" for us to enter the kingdom (10:17–22; compare 4:19); but even that sequence has some of its sting drawn. "Hard" it may be for the rich to enter the kingdom, yet "all things are possible with God" (10:23, 25, 27). Indeed, the conclusion to that particular conversation implies that willingness to abandon all for the sake of Jesus is *not* followed by a life without human ties, even "now in this time," but rather by its opposite (10:30). While it may be conceded that this passage in particular refers to the believer's new "family" in the Church (compare 3:31–35), still the followers of Jesus in Mark are made powerfully aware that ordinary human marriage remains a lifelong commitment, precious in God's sight (10:1–12), and that children, the natural fruit of marriage, are not to be "hindered" (10:14; probably a baptismal phrase: compare Acts 8:36, 10:47) in their relationship with Jesus. The Markan community, like all early Christian communities, knows the danger of persecution, and must be emboldened to meet it (13:9–13); but it is stable communities with ties, not rootless charismatics, who are most vulnerable to such persecution by the state, and it is in fact a community with such ties that is warned by Mark of the divisions and dangers that persecution will bring (13:12).

Of course the ancient world was used to wandering teachers and seers, and Jesus' followers as they set off on their mission to change the world no doubt resembled such teachers (6:7–12). The mere fact that Jesus' disciples, like (say) the Cynics, generally aimed their message at ordinary members of the populace would lead to such resemblance:

> . . . we are bound to assume that the presuppositions, the prior interests of those who joined in, and those being approached, are likely to have had a significant effect on what was said and later written. At the very least, the listeners will have made clear what was getting home to them, and what was leaving them cold, what was answering their questions and what seemed to be beside the point, and not worth repeating. This will have affected selection; it may also of course have influenced style, and even content (Downing, *Christ and the Cynics*, v).

(Consider, by way of analogy, the evident similarities of style and language between modern American television evangelism and television advertisements for automobiles, medicines, and insurance!) Clearly, moreover, Christians and Cynics really did have a number of themes in common, such as a dislike of cant and hypocrisy, and a love of simplicity. Yet when we have granted all this, and when we have further conceded that Cynics, like Christians, often differed among themselves, yet it remains that Cynics had a certain popular image, and that in a number of respects Jesus' disciples as Mark presents them differ from that image. Unlike Cynics who sneered at hospitality and tended deliberately to abuse those who offered kindness (Diogenes Laertius, 6.26, 32, 34), Jesus' disciples are to accept whatever hospitality is offered

(6:10; compare 9:41). Unlike Cynics whose characteristics were a ragged philosophers' cloak, bare feet, no shirt, and a beggars' wallet (πήρα) (Lucian, *The Cynic* 1; *Dialogues of the Dead* 331, 378; Diogenes Laertius 6.23, 31, 33), the disciples of Jesus will go shod, will wear a (single) tunic, and will have no wallet (πήρα: RSV, "bag"); perhaps most significant of all, the cloak is not even mentioned (6:8–9). Leaving aside more important distinctions (such as the Christian proclamation of the kingdom of God and of Jesus as the Christ) I doubt whether even these comparatively superficial differences would have gone unnoticed by Mark's audience.

Mark the Evangelist and the First Urban Christians

So much may be said negatively with regard to Kee's arguments. More positively, my entire impression of Mark's "Life of Jesus" and the processes that will have gone into its making is that it is a somewhat bourgeois production (I use the word in no pejorative sense). While it is certainly not a product of high culture, I see the mindset that envisaged it, the situation that produced it, and above all the situation that could have made use of it, as much more likely to have existed amid the comparative domestic stability and economic prosperity of those disciples whom Wayne Meeks has characterized as "the first urban Christians," than among wandering charismatics. In Mark, "not the temple, not the tomb, but the house is the center of Jesus' action," suggests Elizabeth Struthers Malbon (*Narrative Space*, 117); and we might add, not the street corner, and not the open road. Much of Jesus' teaching, and often the most important teaching, is given "at home" (*en oikō*(*i*)), or "in the house" (εἰς τὴν οἰκίαν) (compare 2:1, 7:17–23, 24–30, 9:33, 10:10–12), and Jesus is shown both as host and as guest (1:29, 2:15, 3:20, 14:12–31). Moreover, "a house in Mark is no place for secrets, for even when Jesus 'would not have any one know' of his presence in a house, 'he could not be hid' (7:24)" (Malbon, 116).

I have just spoken of the "comparative" stability of Markan disciples, and the word "comparative" is important. It is also clear that underlying that stability is a great instability, though it is not the instability of those who choose to be wandering charismatics. We have noted similarities of style that link Mark with other popular literature of the period, with literature such as the Greek romances addressed, as B. P. Reardon has put it, to audiences "fairly wide and not unduly cultivated" (*Form of Greek Romance*, 41). We should further note that for all that separates Mark from the romances, the two have something in common besides style. In terms of their outlook upon the world, both reflect the preoccupations of those who have existed for generations within the confines of empire. To quote Reardon again:

> While local life was assuredly busy, in the wider context of the cosmopolitan Greek world this was no longer the compact culture . . . in which a man could aspire to having an effective voice in controlling his own social existence in his own autonomous community; now there was a large-scale, open society, in which the individual cut a much smaller figure, was swallowed up and lost in the mass—as we may feel lost in today's large-scale open society. Chariton's story [of the love affair of Callirhoe and Chaereas] was written for those who lived in that world, and it reflects that world and its inhabitants, their situation, their anxieties, their aspirations.
>
> In short, this is fable, *mythos*; myth, even. This narrative expresses a social and personal myth, of the private individual isolated and insecure in a world too big for him, and finding his security, his very identity, in love. Chaereas, well-born and wealthy as he is, is nonetheless a nobody, on the world stage. He is not one of those who make the world move, a king of Persia, a Dionysius, a Mithridates (28–29).

What then of Mark's audience? They too, apparently, are not those who make the world move. They, like Chaereas and Callirhoe, may find themselves dragged before kings and governors. They, like Chaereas and Callirhoe, may find themselves the victims of others' violence. Mark's narrative reflects (to paraphrase Reardon) the concerns and anxieties of private individuals isolated and insecure in a world too big for them, who find their security and their very identity in the Christian community and in the fellowship of Jesus Christ.

BIBLIOGRAPHY

On Mark in performance, see Janet Karsten Larson, "St. Alec's Gospel," *Christian Century* 96.1 (1979): 17–19; also John Koenig, "St. Mark on the Stage: Laughing all the Way to the Cross," *Theology Today* 36.1 (1979): 84–86 (Koenig liked McCowen's interpretation a good deal less than did

Larson, but is in no doubt about Mark's suitability for dramatic production). See the bibliography for Chapter 5.

On the presence of oral characteristics in Old English poetry even when it was composed in writing, see Larry Benson, "The Literary Character of Anglo-Saxon Formulaic Poetry," *PMLA* 81 (1966): 334–41; for a development of Benson's position, see Michael Zwettler, *The Oral Tradition of Classical Arabic Poetry: Its Character and Implications* (Columbus: Ohio State University Press, 1978), 15–19; compare H. J. Chaytor, *From Script to Print: An Introduction to Medieval Vernacular Literature* (New York: October House, 1967), 10–13. On the *Shâhnâma*, see O. M. Davidson, "A Formulaic Analysis of Samples Taken from the *Shâhnâma* of Firdowsi," *Oral Tradition* 3.1–2 (1988): 88–105.

On notebooks and notetaking in antiquity, see Colin H. Roberts and T. C. Skeat, *The Birth of the Codex* (London: Oxford University Press for The British Academy, 1987), especially 11–23; there are also general comments and bibliography on the creation of documents in late antiquity in Paul J. Achtemeier, "*Omne verbum sonat*: The New Testament and the Oral Environment of Late Western Antiquity," *JBL* 109.1 (1990): especially 11–15.

For discussion of variations in the Markan textual tradition see Helmut Koester, *Ancient Christian Gospels: Their History and Development* (Philadelphia: Trinity, 1990), 275–86 and literature there cited; for the evidence of *Secret Mark*, see ibid., 293–303. I of course, do not necessarily agree with all Koester's opinions. I doubt very much whether Mark at any stage lacked 6:45–8:28, save possibly through the accidental loss of pages; on which possibility, see Ernst Haenchen, *Der Weg Jesu* (Berlin: Töpelmann, 1966), 303–4 (cited in Koester, 284).

With regard to the possibility of a "scribal" Mark, see Mary Ann Beavis, *Mark's Audience: The Literary and Social Setting of Mark 4:11–12* (Sheffield: Sheffield Academic, 1989).

On the view that the Markan community were wandering charismatics, see Howard Clark Kee, *Community of the New Age: Studies in Mark's Gospel* (Philadelphia: Westminster, 1977). Resemblances and differences between the Jesus and the Cynic traditions are splendidly displayed in F. Gerald Downing, *Christ and the Cynics: Jesus and Other Radical Preachers in First-Century Tradition* (Manuals 4, Sheffield, JSOT, 1988), though at the end of the day I still find the differences, at least as far as Mark is concerned, considerably more striking than the resemblances.

For the "urban" background and audience for Mark suggested above, see Wayne A. Meeks, *The First Urban Christians: The Social World of the Apostle Paul.* (New Haven: Yale University Press, 1983); also Vincent Branick, *The House Church in the Writings of Paul* (Wilmington, Del.: Glazier, 1989). The background and intended audience of the Greek romances, like that of the

New Testament, has been the subject of speculation in recent decades. Foundational is Ben Edwin Perry, *The Ancient Romances: A Literary-Historical Account of their Origins* (Berkeley and Los Angeles: University of California Press, 1967); see further B. P. Reardon, "Aspects of the Greek Novel," *Greece and Rome*, 2d ser., 23.2 (1976): 129–31 and Thomas Hägg, *The Novel in Antiquity* (Berkeley and Los Angeles: University of California Press, 1983), especially 81–108. It is already quite clear that those interested in the setting of the New Testament and those interested in the setting of Greek popular literature in general will do well to keep an eye on each other.

13

Unscientific Postscripts

Many Traditions and One Gospel

A recurring feature of religious history is the way in which groups of persons who hold widely differing, or even opposed, emphases and beliefs, will yet see themselves as the legitimate heirs of a single figure. Those who claim to be heirs of St. Paul, Mahomet, and Martin Luther could all be offered as examples. Even with a figure so close to us in time and so well documented as C. S. Lewis, we find already the development of a "Catholic" Lewis, endorsed by Walter Hooper and the C. S. Lewis Society in Oxford, and an "Evangelical" Lewis endorsed by the Marion E. Wade Center at Wheaton College, Illinois. It is not here our concern to discuss the reason for such a phenomenon (which, incidentally, in the case of Lewis has been well discussed by A. N. Wilson in a recent biography), but simply to note that we need hardly be surprised to find exactly the same thing happening in the case of Jesus of Nazareth. I refer, not to the comparatively recent division of Christendom into its major groupings, but to variations of emphasis and understanding that seem to have been present almost from the beginning.

Helmut Koester and others have argued that various traditions about Jesus competed in the apostolic period, and left their marks on our sources (see Koester, "One Jesus," *Introduction to the New Testament*, and *Ancient Christian Gospels*). One tradition saw Jesus' words as prophetic: heavenly Wisdom summoning the community of the new age to behavior that demonstrated the presence of the kingdom. This tradition is represented in the first stage of the (hypothetical) synoptic source "Q," and in the *Gospel of Thomas*. Another concentrated on a Jesus who spoke words of apocalyptic prophecy, revealing things to come and the immediate coming of the Son of man. Originating perhaps in apocalyptic sayings such as those quoted by Paul in 1 Thessalonians 4:15, this tradition is represented in its more developed form by the canonical Revelation to John, and by the apocalyptic section of the *Didache* (16:1–8). A third tradition, directly opposed by Paul in 2 Corinthians, appears to have focused on Jesus the divine man and worker of miracles, whose supernatural power showed that he was Son of God. Against this, Paul opposed his own tradition: the kerygma of the cross and resurrection, wherein a major focus was the understanding of Jesus' death and vindication as in accordance with Scripture, a tradition we find linked not only with Paul (1 Cor. 1:23, 15:1–8), but also with Peter (Gal. 1:18; compare 1 Peter 2:21–24, *GosPet.* passim; *Kerygma Petrou* [Dobschütz, fragment 9; Schneemelcher, 102]).

Mark's gospel contains material representative of all these views. As regards Jesus the teacher, there has been a tendency among critics to regard this element as rather unimportant in Mark, and certainly Mark contains less teaching material than do Matthew or Luke. Yet Mark does bring Jesus onto the scene from the start as one who proclaims the kingdom of God, and calls us to repentence (1:15); in 4:1–34 we have words of Jesus that are clearly presented as heavenly wisdom revealing the "mystery of the kingdom of God" to disciples; and we do find in Mark both the expectation of the Son of man in his "Day"—which seems to have been a feature of the revision stage of "Q" (Matt.; compare Mark 8:38)—and the teaching of the presence of the kingdom for the believer that dominates the *Gospel of Thomas* (for example, *GosTh.* 3, 22, 27 etc.; compare Mark 1:15), although in Mark it is balanced by teaching about the future presence of the kingdom (for example, 9:1). The Jesus of apocalyptic prophecy is represented with particular clarity in the Farewell Discourse (Mark 13), where Jesus speaks of the Fall of Jerusalem and the persecution of the Church, as well as of the future coming of the Son of man. Jesus the worker of miracles is to be found throughout the gospel, but perhaps especially in 4:35–5:43, where he shows himself Lord of the elements, the demons, disease, and death. Lastly, of course, Mark's gospel is completed and indeed dominated by the kerygma of the cross and resurrection (8:31, 9:31, 10:33–34, 14–16). Since that kerygma is, as we have just observed, particularly associated with Peter and Paul, it is interesting to note that there are venerable traditions associating Mark with both apostles (1 Peter 5:13; Eusebius, *Ecclesiastical History* 3.39.15–16; Philem. 24; compare Col. 4:10, 2 Tim. 4:11, Acts 12:25, 13:13, 15:37).

Nonetheless, Mark has achieved something that Paul did not achieve. Paul simply opposed his preaching of the cross to what he regarded as the false wisdom of those who preached a Jesus known only after the flesh. His was only the realization of one perspective within a tradition that did, after all, whether Paul liked it or not (and he clearly didn't), allow other perspectives. Mark presents the two views in a creative tension. Indeed, he offers to us not only the Christ of the mighty works, but also Jesus the teacher of God's kingdom, and Jesus the eschatological prophet, all in the light of Jesus crucified and risen. If, moreover, as seems likely, Mark wrote in the immediate aftermath of the Jewish War (or even if he wrote during it) then we should be foolish to underestimate the degree of political resonance that there would have been for him, and presumably for his hearers, in the titles and roles the tradition claimed for Jesus: "the anointed," "the son of David," and the proclaimer of "God's kingdom." (As a simple illustration of the degree of political danger implicit in the title "Son of David," we may note that following the conquest of Jerusalem, Vespasian ordered that all descendants of David should be sought out, "so that, among the Jews, not one of the royal house should remain," and that as late as the reigns of Domitian and Trajan, the Roman administration was still concerned about Davidic descendants [Eusebius, *Ecclesiastical History* 3.12.19, 32]). Nonetheless, these titles and roles, too, Mark does not hesitate to claim for Jesus, while making clear that, unlike others who had claimed them, Jesus' claim and Jesus' proclamation involved a willingness to accept freely only one death, his own.

All this Mark has achieved, in the first instance, by his willingness to bring together all the "matter" of Jesus—all the ways in which different persons and groups had remembered and revered him—in a single "Life." Moreover, the gospel's use of the techniques of oral storytelling, such as its repetition of the motifs of Jesus' authority, his closeness to God, and his victory over Satan, coupled with (in the first part of the narrative) the query cast over Jesus' mighty acts, and even over naming him for who he is, by repeated injunctions to silence and occasional hints of the Passion, and (in the second part of the narrative) by repeated anticipations of the cross and of the coming triumph of the Son of man—all these press the hearers not to hold to the Son of God without realizing that his way leads to a cross (8:34), nor to hold to the crucified without realizing who and Whose he is (15:39).[1]

The Significance of the Story

In organizing the "matter" of Jesus of Nazareth as a "life," Mark's text imposes upon it a narrative form. Without such an organization, Christian preaching and Christian writing might have continued merely as proposition, claim, or confession, interpreting and to be interpreted in the light of Scripture. By presenting the gospel as story, Mark took an important step along the path that was to lead to the establishment of the New Testament alongside the Old as Scripture in its own right, requiring in its turn to be interpreted: in Judeo-Christian tradition it is the *story*, essentially, that foments faith, and then (in the light of that faith) leads to theological interpretation.

The Old Testament, as Paul Ricouer has pointed out, can be summed up in the question, "Who is the Lord?" ("Interpretative Narrative"). The New, as Hans Frie suggested, centers upon "Who do you say that I am?"—the identity of Jesus (*The Identity of Jesus Christ*). Old and New face those questions—or rather, demand that we face them—primarily by offering narrative: God's dealing with Israel, and the "life" of Jesus. Of course confession and proposition can answer such questions in their own way. What they cannot do is involve us directly in the pain and ambiguity involved in any answer offered in that realm of history where, as Robert Alter puts it, "in the stubbornness of human individuality . . . each man and woman encounters God or ignores Him, responds to or resists Him" (*Art of Biblical Narrative*, 189). Only narrative can lead us to identify with Jacob in his struggle with the angel, and with God's pathos over the disobedience of Israel; only narrative can lead us to identify with the disciples' constant failures, and with

1. There is in this connection an obvious analogy between the activity of Mark and that of the Fourth Evangelist, who, as Helmut Koester has recently pointed out, "used the passion narrative to interpret the gnosticizing traditions of his own community and has brought this community and its special heritage into the movement of Panchristianity" ("Story of the Johannine Tradition," 32).

the Son of God's grief in Gethsemane; only narrative can involve us in the promise given at the Last Supper and the treachery simultaneously planned; only narrative can lead us to identify with the terrible ambiguity of the cross and with the final, awestruck silence of the women at the tomb. All these points of narrative are, of course, obscure. They are often negative, and generally confusing. At an earlier stage in this study we observed the deep connection that exists between narrative and oral ways of knowing; the truth is that narrative burdens us with that emotional involvement characteristic of oral knowing. But all this—the negativity, the confusion, and the emotional involvement—is also characteristic of life itself as we experience it. It is part of the way things are in the "perilously momentous realm of history" (Alter, *Biblical Narrative*, 189). The Platonic demand for clear reasoning represents something that is vital for us if we are to achieve maturity; but it does not represent the only thing. The ancients continued to value Homer as well as Plato, and they were right to do so.

Essentially, therefore, Rudolf Bultmann (*Synoptic Tradition*) was correct in his assertion that there was something in the nature of the gospel that required its being remembered as the gospel story, culminating in the kerygma of cross and resurrection. There is, in Ricoeur's phrase ("Interpretative Narrative," 242), a "deep-lying affinity" linking the memory of Jesus' sharing the table of sinners and disciples, his teaching, his healing, his "handing over" (*paradōis*) to the authorities for suffering and death, his being raised from the dead, and his continued "handing over" to the Church in the proclamation of the gospel. Mark's gift to the Church was (under God) that he used the ambiguity, negativity, and confusion of a story to proclaim the crucified and risen Son of God.

Gospel and Performance

Again and again in our reflections on Mark we have found ourselves face-to-face with the issue of performance. As we have observed, in an age without printing, performance was inevitable if there was to be anything approaching publication; yet we should be wrong to imagine that for the ancients, any more than for ourselves, performance was merely an expedient. Quintilian was aware, certainly, that for the exercise of some kinds of critical faculty, private reading and private study were helpful; yet he was also aware that quite different advantages were conferred by listening to a text: "Speakers stimulate us by the animation of their delivery, and kindle the imagination, not by presenting us with an elaborate picture, but by bringing us into actual touch with the things themselves" (*Institutio Oratoria* 10.1.16, trans. H. E. Butler, 1933, adapted). What happens when a text is performed, rather than read or studied in silence? First, and fundamentally, words are returned to their native element (Ong, *Presence of the Word*, 1–35). Especially is this true of words that were written to be heard. But of performance (as opposed, say, to conversation) we must say more. In performance, three elements are involved, and three persons or groups of persons. First, there is a text, which an author has created; second, there is an interpretation by the performer or performers; and third, there is the experience of that performance with an audience. Alla Bozarth-Campbell, adapting as her conceptual model Christian understandings of the Incarnation, has some interesting observations about the significance of this:

> First there is mutual *in*forming: the form or image of the word flows into the form of the interpreter, and the interpreter . . . begins to put bodily form into it. . . . The image form of the text is changed into bodily presence, and the physical form of the interpreter/actor is changed into the icon/servant of the word. All of this change comes about in *per*formance, as the word's form is brought to fulfillment through the interpreter/actor in the whole process that is finally seen and heard in the *kairos* [moment] of incarnation. . . . (*Word's Body*, 141).
>
> The *becoming* process of encounter and coalescence between word and interpreter is consummated in the revelation of the poem's *being* through the interpreter as icon in the transforming moment of performance. . . . The world of the poem is delivered from the silence and stillness of the word in potential and is revealed as it expresses itself through the interpreter as physical and spiritual icon. Through the loving cooperation of the interpreter the word is received into the lives of those who see and hear it in performance. Audience and interpreter participate in the world of the poem and act together in the ambience of its vitality. (143)

Such an understanding of performance makes considerable demands of performers and, if the text is important to us, places heavy responsibility upon them. They must be prepared, in the deepest sense of which they are capable, to "know" the text. They must be willing to sit under its judgment. Here, too, in its degree, the response "Be it unto me according to thy word" is required. This is something that the best performers have always known. Konstantin Stanislavsky described the necessary work of an actor as "the study of the spiritual essence of a dramatic work, the germ from which it has emerged and which defines its meaning as well as the meaning of its parts. The worst enemy of progress is prejudice" (cited in David Magarshack, *Stanislavsky*, 27). By what other means could one create an icon?

"What," asks one of my students, with reference to a difficult gospel parable, "was this stuff actually *like*? Was it funny? Angry? Pleading? Judicial?" I cannot answer her. We have only the text. The word written is a substitute for presence, and no more than that. Yet sympathetic performance by a performer can create an icon. Many performances can create many icons. And these icons can show us more of those originals than even their creators may have recognized.

The Gospel according to Mark performed by twentieth-century performers for a twentieth-century audience will not be the same as that performed in the first century for the first-century audience. Even on the same day in the twentieth century, it will not be the same performed in Europe or North America as it will be in South America or India. Any performers worth their salt will tell us that if two performances of the same text are given by the same company at the same place on successive nights to audiences that to all intents and purposes are socially and culturally identical, those two performances will be different from each other. Every performance is unique. But every performance of an author's work is an icon created with the author in combination with interpreters and audience, and that is the most that any author could or can ever expect. The result, moreover, is an experience of the author, as the result of Laurence Olivier's and Kenneth Branagh's versions of *Henry V* is in both cases an experience of Shakespeare, although in each case a different Shakespeare, different from each other, different from the Shakespeare encountered by those who first sat south of the Thames and watched Shakespeare's own players attempt to cram within a wooden O the very casques that did affright the air at Agincourt, and different from the Shakespeare encountered in every other performance of *Henry V* that has ever been. But Shakespeare, nonetheless. It is thus that an author who has written for performance continues to live and be in conversation with us; and it is thus, I believe, that we must begin again to use Mark (and, indeed, all ancient texts) if we wish to share the experience that they would have given to those who first encountered them. This is equally true, whether our concern is merely antiquarian and historical, or whether we approach as believers.

I have written elsewhere ("Preachers and Critics"; compare "The Judicious Mr. Hooker," 158) of the abyss created between ourselves and ancient texts by the questions which our Enlightenment inheritance obliges us to ask. We have therefore examined Mark's text using the tools of literary and historical criticism. If nothing else becomes clear from such an examination, it is surely that Mark's literary and social milieu is very remote from ours, and that there is much about it we shall probably never understand. The abyss that separates Mark from us can, however, be bridged by one means: performance. Even performance, of course, will not enable me to answer my student's question; I cannot know what this "stuff" actually was like for those who first heard it. What we can both discover is what the stuff is like for us, at a particular moment, in a particular setting. Even from Mark's point of view, perhaps that would have been what mattered.

BIBLIOGRAPHY

On varying early traditions about Jesus, see Helmut Koester, "One Jesus and Four Primitive Gospels," in James M. Robinson and Helmut Koester, *Trajectories through Early Christianity* (Philadelphia: Fortress, 1971), 158–204; also Koester's *Introduction to the New Testament,* vol. 2: *History and Literature of Early Christianity* (Berlin: Walter de Gruyter, 1982). On Q, again see Koester's *Ancient Christian Gospels: Their History and Development* (London: SCM; Philadelphia: Trinity, 1990), 128–72, and for a somewhat different view, John S. Kloppenborg, *The Formation of Q: Trajectories in Ancient Wisdom Collections* (Philadelphia: Fortress, 1987).

On the opponents of Paul, see Dieter Georgi, *The Opponents of Paul in 2 Corinthians* (Philadelphia: Fortress, 1986).

On the theological significance of narrative, there is a vast literature that I do not pretend to have mastered. The following, however, have been found suggestive: Eric Auerbach, *Mimesis: The Representation of Reality in Western Literature*, trans. Willard R. Trask (Princeton: Princeton University Press, 1953); Robert Alter, *The Art of Biblical Narrative* (New York: Basic Books, 1981); Frank Kermode, *The Genesis of Secrecy: On the Interpretation of*

Narrative (Cambridge: Harvard University Press, 1979); Paul Ricoeur, "Interpretative Narrative," in Regina Schwartz, ed., *The Book and the Text: The Bible and Literary Theory* (Oxford: Basil Blackwell, 1990); Hans Frei, *The Identity of Jesus Christ: An Enquiry into the Hermeneutical Basis of Dogma* (Philadelphia: Fortress, 1974).

On the significance of performance, see Alla Bozarth-Campbell, *The Word's Body: An Incarnational Aesthetic of Interpretation* (Tuscaloosa: University of Alabama Press, 1979); also suggestive is Dorothy Sayers's *The Mind of the Maker* (London: Methuen, 1941). See also Paul Zumthor, *Oral Poetry: An Introduction* (Minneapolis: University of Minnesota, 1990), especially 117–229.

Appendix

Examples of Popular Greek Prose from the First and Second Centuries of the Christian Era

At present the relevant texts are not always easy to obtain. Readers who wish to try for themselves the experiment suggested on page 54 above may therefore find the following extracts useful.

1. An extract from *The Life of Secundus the Philosopher*: text and translation from Ben Edwin Perry, *Secundus the Silent Philosopher*, APA Monographs 22 (Ithaca: American Philological Association and Cornell University Press).

ΒΙΟΣ ΣΕΚΟΥΝΔΟΥ ΦΙΛΟΣΟΦΟΥ

Σεκοῦνδος ἐγένετο φιλόσοφος. οὗτος ἐφιλοσόφησε τὸν ἅπαντα χρόνον σιωπὴν ἀσκήσας, Πυθαγορικὸν ἐξειληφὼς βίον. τὸ δ' αἴτιον τῆς σιωπῆς τοῦτο· ἐπέμφθη παρὰ τῶν γονέων μικρὸς ὢν παιδευθῆναι. ὄντος δὲ αὐτοῦ ἐν τῇ παιδεύσει ἐγένετο τὸν πατέρα αὐτοῦ τελευτῆσαι. ἦν δὲ ἀκούων περὶ τῆς παραβολῆς ταύτης· ὅτι πᾶσα γυνὴ πόρνη, ἡ δὲ λαθοῦσα σώφρων. τέλειος οὖν γενάμενος ἐπανῆλθεν εἰς τὴν ἰδίαν πατρίδα, τὴν τοῦ κυνὸς προφέρων ἄσκησιν· βάκλον καὶ πήραν περιφέρων, τὴν κεφαλὴν καὶ τὸν πώγωνα ἀναθρέψας. λαμβάνει οὗτος μετάτον ἐν τῇ ἰδίᾳ οἰκίᾳ, μηδενὸς τῶν οἰκείων αὐτὸν γνωρίζοντος, μηδὲ τῆς ἰδίας μητρός. βουλόμενος δὲ πεισθῆναι καὶ τὸν περὶ γυναικῶν λόγον, εἰ ἄρα ἀληθής ἐστι, καλέσας μίαν τῶν παιδισκῶν ὑπέσχετο αὐτῇ παρασχεῖν χρυσίνους ἕξ, ὑποκρινόμενος φιλεῖν τὴν κυρίαν αὐτῆς, ἑαυτοῦ δὲ μητέρα. ἡ δὲ λαβοῦσα τὸ χρυσίον ἠδυνήθη πεῖσαι τὴν ἑαυτῆς κυρίαν, ὑποσχομένη αὐτῇ χρυσίνους ν̄. ἡ δὲ συνέθετο τῇ παιδίσκῃ εἰποῦσα ὅτι "ὀψίας ποιήσω αὐτὸν εἰσελθεῖν λάθρα καὶ κοιμηθήσομαι μετ' αὐτοῦ." ὁ δὲ φιλόσοφος ἔχων τὰς ἐπαγγελίας παρὰ τῆς παιδίσκης ἔπεμψεν τὰ πρὸς δεῖπνον. καὶ δὴ τούτων ἀποδειπνησάντων, ὡς ἦλθον πρὸς τὸν ὕπνον, αὕτη μὲν ⟨ἦν⟩ προσδοκῶσα σαρκικῶς αὐτῷ συμμιγῆναι, αὐτὸς δὲ ὡς ἰδίαν μητέρα περιλαμβάνων, καὶ τοῖς ὀφθαλμοῖς περιλάμπων οὓς ἐθήλασε μασθούς, ἐκοιμήθη ἕως πρωΐ. περὶ δὲ τὸ διάφαυσμα ἀναστὰς Σεκοῦνδος ἐβουλεύετο ἐξελθεῖν. ἡ δὲ ἐκράτησεν αὐτὸν λέγουσα· "καταγνῶναί μου θέλων τοῦτο ἐποίησας;" ὁ δὲ εἶπεν· "οὐχί, κυρία μῆτερ, οὐ γὰρ δίκαιόν ἐστιν ὅπερ ἐξῆλθον μιᾶναι· μὴ γένοιτο." ἡ δὲ ἐπυνθάνετο παρ' αὐτοῦ τίς ἂν εἴη. ὁ δὲ εἶπεν αὐτῇ· "ἐγώ εἰμι Σεκοῦνδος, ὁ υἱός σου." ἡ δὲ καταγνοῦσα ἑαυτῆς καὶ μὴ φέρουσα τὴν αἰσχύνην ἀγχόνῃ ἐχρήσατο. ὁ δὲ Σεκοῦνδος, γνοὺς ὅτι διὰ τῆς αὐτοῦ γλώττης ὁ θάνατος τῆς μητρὸς ἐγένετο, ἀπόφασιν καθ' ἑαυτοῦ ἔδωκεν τοῦ μὴ λαλῆσαι τοῦ λοιποῦ· καὶ μέχρι θανάτου τὴν σιωπὴν ἤσκησεν. .·.

[Secundus was a philosopher. This man cultivated wisdom all his days and observed silence religiously, having chosen the Pythagorean way of life. What caused his silence was this: When he was a small boy he was sent away from home by his parents to be educated, and, while he was still occupied with his studies, it happened that his father died. He had often heard this byword, that "every woman can be bought; the chaste one is only she who has escaped notice." Now when he had grown to manhood and had returned to his homeland, he put himself forward as a follower of the Cynic discipline, carrying a stick and a leathern wallet about with him, letting the hair on his head grow long, and cultivating a beard. He took an apartment in his own father's house, without any of the servants recognizing him, nor even his own mother. Wishing to satisfy himself concerning that proposition about women, to see whether it was really true or not, he called one of the maid-servants aside and promised to give her six gold pieces (if she would arrange a meeting for him) on the pretense that he was in love with her mistress, his own mother. The maid, accepting the money, was able to persuade her mistress by promising her fifty gold pieces; and the latter made an agreement with the maid, saying, "At nightfall I will have him enter secretly, and I will lie with him." Having this notice from the maid, the philosopher sent provisions for a dinner. Now after the two had finished dinner, and when they had started to go to bed, she was expecting to have carnal intercourse with him; but he put his arms around her as he would around his own mother, and, fixing his eyes upon the breasts that had suckled him, he lay down and slept until early morning. When the first light of dawn appeared Secundus rose up with the intention of going out, but she laid hands on him and said, "Did you do this only in order to convict me?" And he answered, "No, lady mother, I refrained because it is not right for me to defile that place from which I came forth at birth. God forbid." Then she asked him who he was, and he said to her, "I am Secundus, your son." And she, condemning herself and unable to bear the sense of shame, hanged herself. Secundus, having concluded that it was on account of his own talking that his mother's death had come about, put a ban upon himself, resolving not to say anything the rest of his life. And he practiced silence to the day of his death . . .]

2. An extract from Xenophon of Ephesus, *An Ephesian Tale*: text from Antonius D. Papanikolaou, *Xenophontis Ephesii: Ephesiacorum Libri V de Amoribus Anthiae et Abrocomae* (Leipzig: Teubner, 1973):

English translation from B. P. Reardon, ed., *Collected Ancient Greek Novels* (Berkeley and Los Angeles: University of California Press, 1989).

ΞΕΝΟΦΩΝΤΟΣ ΤΩΝ ΚΑΤΑ ΑΝΘΙΑΝ ΚΑΙ ΑΒΡΟΚΟΜΗΝ ΕΦΕΣΙΑΚΩΝ ΛΟΓΟΣ ΠΡΩΤΟΣ

Ἦν ἐν Ἐφέσῳ ἀνὴρ τῶν τὰ πρῶτα ἐκεῖ δυναμένων, Λυκομήδης ὄνομα. τούτῳ τῷ Λυκομήδει ἐκ γυναικὸς ἐπιχωρίας Θεμιστοῦς γίνεται παῖς Ἁβροκόμης, μέγα δέ τι χρῆμα [ὡραιότητι σώματος ὑπερβαλλούσῃ] κάλλους οὔτε ἐν Ἰωνίᾳ οὔτε ἐν ἄλλῃ γῇ πρότερον γενομένου. οὗτος ὁ Ἁβροκόμης ἀεὶ μὲν καὶ καθ' ἡμέραν εἰς κάλλος ηὔξετο, συνήνθει δὲ αὐτῷ τοῖς τοῦ σώματος καλοῖς καὶ τὰ τῆς ψυχῆς ἀγαθά· παιδείαν τε γὰρ πᾶσαν ἐμελέτα καὶ μουσικὴν ποικίλην ἤσκει, καὶ θήρα δὲ αὐτῷ καὶ ἱππασία καὶ ὁπλομαχία συνήθη γυμνάσματα. ἦν δὲ περισπούδαστος ἅπασιν Ἐφεσίοις, ἀλλὰ καὶ τοῖς τὴν ἄλλην Ἀσίαν οἰκοῦσι, καὶ μεγάλας εἶχον ἐν αὐτῷ τὰς ἐλπίδας ὅτι πολίτης ἔσοιτο διαφέρων. προσεῖχον δὲ ὡς θεῷ τῷ μειρακίῳ· καί εἰσιν ἤδη τινὲς οἳ καὶ προσεκύνησαν ἰδόντες καὶ προσηύξαντο. ἐφρόνει δὲ τὸ μειράκιον ἐφ' ἑαυτῷ μεγάλα καὶ ἠγάλλετο μὲν καὶ τοῖς τῆς ψυχῆς κατορθώμασι, πολὺ δὲ μᾶλλον τῷ κάλλει τοῦ σώματος· πάντων δὲ τῶν ἄλλων, ὅσα δὴ ἐλέγετο καλά, ὡς ἐλαττόνων κατεφρόνει καὶ οὐδὲν αὐτῷ, οὐ θέαμα, οὐκ ἄκουσμα ἄξιον Ἁβροκόμου κατεφαίνετο· καὶ εἴ τινα ἢ παῖδα καλὸν ἀκούσαι ἢ παρθένον εὔμορφον, κατεγέλα τῶν λεγόντων ὡς οὐκ εἰδότων ὅτι εἷς καλὸς αὐτός. Ἔρωτά γε μὴν οὐδὲ ἐνόμιζεν εἶναι θεόν, ἀλλὰ πάντῃ ἐξέβαλεν ὡς οὐδὲν ἡγούμενος, λέγων ὡς οὐκ ἄν ποτε οὐ⟨δὲ⟩ εἷς ἐρασθείη οὐδὲ ὑποταγείη τῷ θεῷ μὴ θέλων· εἰ δέ που ἱερὸν ἢ ἄγαλμα Ἔρωτος εἶδε, κατεγέλα, ἀπέφαινέ τε ἑαυτὸν Ἔρωτος παντὸς καλλίονα καὶ κάλλει σώματος καὶ δυνάμει. καὶ εἶχεν οὕτως· ὅπου γὰρ Ἁβροκόμης ὀφθείη, οὔτε ἄγαλμα ⟨καλὸν⟩ κατεφαίνετο οὔτε εἰκὼν ἐπῃνεῖτο.

Μηνιᾷ πρὸς ταῦτα ὁ Ἔρως· φιλόνεικος γὰρ ὁ θεὸς καὶ ὑπερηφάνοις ἀπαραίτητος· ἐζήτει δὲ τέχνην κατὰ τοῦ μει-

*ρακίου· καὶ γὰρ καὶ τῷ θεῷ δυσάλωτος ἐφαίνετο. ἐξοπλίσας οὖν ἑαυτὸν καὶ πᾶσαν δύναμιν ἐρωτικῶν φαρμάκων περιβαλόμενος ἐστράτευεν ἐπ' Ἀβροκόμην. ἤγετο δὲ τῆς Ἀρτέμιδος ἐπιχώριος ἑορτὴ ἀπὸ τῆς πόλεως ἐπὶ τὸ ἱερόν· στάδιοι δέ εἰσιν ἑπτά· ἔδει δὲ πομπεύειν πάσας τὰς ἐπιχωρίους παρθένους κεκοσμημένας πολυτελῶς καὶ τοὺς ἐφήβους, ὅσοι τὴν αὐτὴν ἡλικίαν εἶχον τῷ Ἀβροκόμῃ. ἦν δὲ αὐτὸς περὶ τὰ ἓξ καὶ δέκα ἔτη καὶ τῶν ἐφήβων προσήπτετο καὶ ἐν τῇ πομπῇ τὰ πρῶτα ἐφέρετο. πολὺ δὲ πλῆθος ἐπὶ τὴν θέαν, πολὺ μὲν ἐγχώριον, πολὺ δὲ ξενικόν· καὶ γὰρ ἔθος ἦν ⟨ἐν⟩ ἐκείνῃ τῇ πανηγύρει καὶ νυμφίους ταῖς παρθένοις εὑρίσκεσθαι καὶ γυναῖκας τοῖς ἐφήβοις. παρῄεσαν δὲ κατὰ στίχον οἱ πομπεύοντες· πρῶτα μὲν τὰ ἱερὰ καὶ δᾷδες καὶ κανᾶ καὶ θυμιάματα· ἐπὶ τούτοις ἵπποι καὶ κύνες καὶ σκεύη κυνηγετικά, ἔτι καὶ πολεμικά, τὰ δὲ πλεῖστα εἰρηνικά. ** ἑκάστη δὲ αὐτῶν οὕτως ὡς πρὸς ἐραστὴν ἐκεκόσμητο. ἦρχε δὲ τῆς τῶν παρθένων τάξεως Ἀνθία, θυγάτηρ Μεγαμήδους καὶ Εὐίππης, ἐγχωρίων. ἦν δὲ τὸ κάλλος τῆς Ἀνθίας οἷον θαυμάσαι καὶ πολὺ τὰς ἄλλας ὑπερεβάλλετο παρθένους. ἔτη μὲν τεσσαρεσκαίδεκα ἐγεγόνει, ἤνθει δὲ αὐτῆς τὸ σῶμα ἐπ' εὐμορφίᾳ, καὶ ὁ τοῦ σχήματος κόσμος πολὺς εἰς ὥραν συνεβάλλετο· κόμη ξανθή, ἡ πολλὴ καθειμένη, ὀλίγη πεπλεγμένη, πρὸς τὴν τῶν ἀνέμων φορὰν κινουμένη· ὀφθαλμοὶ γοργοί, φαιδροὶ μὲν ὡς κόρης, φοβεροὶ δὲ ὡς σώφρονος· ἐσθὴς χιτὼν ἁλουργής, ζωστὸς εἰς γόνυ, μέχρι βραχιόνων καθειμένος, νεβρὶς περικειμένη, γωρυτὸς ἀνημμένος, τόξα ὅπλα, ἄκοντες φερόμενοι, κύνες ἑπόμενοι. πολλάκις αὐτὴν ἐπὶ τοῦ τεμένους ἰδόντες Ἐφέσιοι προσεκύνησαν ὡς Ἄρτεμιν. καὶ τότ' οὖν ὀφθείσης ἀνεβόησε τὸ πλῆθος, καὶ ἦσαν ποικίλαι παρὰ τῶν θεωμένων φωναί, τῶν μὲν ὑπ' ἐκπλήξεως τὴν θεὸν εἶναι λεγόντων, τῶν δὲ ἄλλην τινὰ ὑπὸ τῆς θεοῦ πεποιημένην· προσηύχοντο δὲ πάντες καὶ προσεκύνουν καὶ τοὺς γονεῖς αὐτῆς ἐμακάριζον· ἦν δὲ διαβόητος τοῖς θεωμένοις ἅπασιν Ἀνθία ἡ καλή. ὡς δὲ παρῆλθε τὸ τῶν παρθένων πλῆθος, οὐδεὶς ἄλλο τι ἢ Ἀνθίαν ἔλεγεν· ὡς δὲ Ἀβροκόμης μετὰ τῶν ἐφήβων ἐπέστη, τοὐνθένδε, καίτοι καλοῦ ὄντος τοῦ κατὰ τὰς παρθένους θεάματος, πάντες ἰδόντες Ἀβροκόμην ἐκείνων ἐπελάθοντο, ἔτρεψαν δὲ τὰς ὄψεις ἐπ' αὐτὸν βοῶντες ἀπὸ τῆς θέας ἐκπεπληγμένοι, ,,καλὸς Ἀβρο-*

κόμης" λέγοντες „καὶ οἷος οὐδὲ εἷς καλοῦ μίμημα θεοῦ". ἤδη δέ τινες καὶ τοῦτο προσέθεσαν· „οἷος ἂν γάμος γένοιτο Ἀβροκόμου καὶ Ἀνθίας".

Καὶ ταῦτα ἦν πρῶτα τῆς Ἔρωτος τέχνης μελετήματα. ταχὺ μὲν δὴ εἰς ἑκατέρους ἡ περὶ ἀλλήλων ἦλθε δόξα· καὶ ἥ τε Ἀνθία τὸν Ἀβροκόμην ἐπεθύμει ἰδεῖν, καὶ ὁ τέως ἀνέραστος Ἀβροκόμης ἤθελεν Ἀνθίαν ἰδεῖν.

Ὡς οὖν ἐτετέλεστο ἡ πομπή, ἦλθον δὲ εἰς τὸ ἱερὸν θύσοντες ἅπαν τὸ πλῆθος καὶ ὁ τῆς πομπῆς κόσμος ἐλέλυτο, ᾔεσαν δὲ ἐς ταὐτὸν ἄνδρες καὶ γυναῖκες, ἔφηβοι καὶ παρθένοι, ἐνταῦθα ὁρῶσιν ἀλλήλους, καὶ ἁλίσκεται Ἀνθία ὑπὸ τοῦ Ἀβροκόμου, ἡττᾶται δὲ ὑπὸ Ἔρωτος Ἀβροκόμης καὶ ἐνεώρα τε συνεχέστερον τῇ κόρῃ καὶ ἀπαλλαγῆναι τῆς ὄψεως ἐθέλων οὐκ ἐδύνατο· κατεῖχε δὲ αὐτὸν ἐγκείμενος ὁ θεός. διέκειτο δὲ καὶ Ἀνθία πονήρως, ὅλοις μὲν καὶ ἀναπεπταμένοις τοῖς ὀφθαλμοῖς τὸ Ἀβροκόμου κάλλος εἰσρέον δεχομένη, ἤδη δὲ καὶ τῶν παρθένοις πρεπόντων καταφρονοῦσα· καὶ γὰρ ἐλάλησεν ἄν τι, ἵνα Ἀβροκόμης ἀκούσῃ, καὶ μέρη τοῦ σώματος ἐγύμνωσεν ἂν τὰ δυνατά, ἵνα Ἀβροκόμης ἴδῃ· ὁ δὲ αὐτὸν ἐδεδώκει πρὸς τὴν θέαν καὶ ἦν αἰχμάλωτος τοῦ θεοῦ.

[Among the most influential citizens of Ephesus was a man called Lycomedes. He and his wife, Themisto, who also belonged to the city, had a son Habrocomes; his good looks were phenomenal, and neither in Ionia nor anywhere else had there ever been anything like them. This Habrocomes grew more handsome every day; and his mental qualities developed along with his physical ones. For he acquired culture of all kinds and practiced a variety of arts; he trained in hunting, riding, and fighting under arms. Everyone in Ephesus sought his company, and in the rest of Asia as well; and they had great hopes that he would have a distinguished position in the city. They treated the boy like a god, and some even prostrated themselves and prayed at the sight of him. He had a high opinion of himself, taking pride in his attainments, and a great deal more in his appearance. Everything that was regarded as beautiful he despised as inferior, and nothing he saw or heard seemed up to his standard. And when he heard a boy or girl praised for their good looks, he laughed at the people making such claims for not knowing that only he himself was handsome. He did not even recognize Eros as a god; he rejected him totally and considered him of no importance, saying that no one would ever fall in love or submit to the god except of his own accord. And whenever he saw a temple or statue of Eros, he used to laugh and claimed that he was more handsome and powerful than any Eros. And that was the case: for wherever Habrocomes appeared, no one admired any statue or praised any picture.

Eros was furious at this, for he is a contentious god and implacable against those who despise him. He looked for some strategem to employ against the boy, for even the god thought he would be difficult to capture. So he armed himself to the teeth, equipped himself with his full armory of love potions, and set out against Habrocomes. The local festival of Artemis was in progress, with its procession from the city to the temple nearly a mile away. All the local girls had to march in procession, richly dressed, as well as all the young men of Habrocomes' age—he was around sixteen, already a member of the Ephebes, and took first place in the procession. There was a great crowd of Ephesians and visitors alike to see the festival, for it was the custom at this festival to find husbands for the girls and wives for the young men. So the procession filed past—first the sacred objects, the torches, the baskets, and the incense; then horses, dogs, hunting equipment . . . some for war, most for peace. And each of the girls was dressed as if to receive a lover. Anthia led the line of girls; she was the daughter of Megamedes and Euippe, both of Ephesus. Anthia's beauty was an object of wonder, far surpassing the other girls'. She was fourteen; her beauty was burgeoning, still more enhanced by the adornment of her dress. Her hair was golden—a little of it plaited, but most hanging loose and blowing in the wind. Her eyes were quick; she had the bright glance of a young girl, and yet the austere look of a virgin. She wore a purple tunic down to the knees, fastened with a girdle and falling loose over her arms, with a fawnskin over it, a quiver attached, and arrows for weapons; she carried javelins and was followed by dogs. Often as they saw her in the sacred enclosure the Ephesians would worship her as Artemis. And so on this occasion too the crowd gave a cheer when they saw her, and there was a whole clamor of exclamations from the spectators: some were amazed and said it was the goddess in person; some that it was someone else made by the goddess in her own image. But all prayed and prostrated themselves and congratulated her parents. "The beautiful Anthia!" was the cry on all the spectators' lips.

When the crowd of girls came past, no one said anything but "Anthia!" But when Habrocomes came in turn with the Ephebes, then, although the spectacle of the women had been a lovely sight, everyone forgot about them and transferred their gaze to him and were smitten at the sight. "Handsome Habrocomes!" they exclaimed, "Incomparable image of a handsome god!" Already some added, "What a match Habrocomes and Anthia would make!"

These were the first machinations in Love's plot. They quickly learned each other's reputation. Anthia longed to see Habrocomes; and Habrocomes, up till now impervious to love, wanted to see Anthia.

And so when the procession was over, the whole crowd went into the temple for the sacrifice, and the files broke up; men and women and girls and boys came together. Then they saw each other, and Anthia was captivated by Habrocomes, while Love got the better of Habrocomes. He kept looking at the girl and in spite of himself could not take his eyes off her. Love held him fast and pressed home his attack. And Anthia too was in a bad way, as she let his appearance sink in, with rapt attention and eyes wide open; and already she paid no attention to modesty: what she said was for Habrocomes to hear, and she revealed what she could of her body for Habrocomes to see. And he was captivated at the sight and was a prisoner of the god.]

3. An extract from Chariton, *Callirhoe*: text from Warren E. Blake, ed., *Charitonis Aphrodisiensis: De Chaerea et Callirhoe Amatoriarum Narrationum Libri Octo* (Oxford: Clarendon, 1938); English from B. P. Reardon, ed., *Collected Ancient Greek Novels* (Berkeley and Los Angeles: University of California Press, 1989).

ΧΑΡΙΤΩΝΟΣ ΑΦΡΟΔΙΣΙΕΩΣ

ΠΕΡΙ ΧΑΙΡΕΑΝ ΚΑΙ ΚΑΛΛΙΡΟΗΝ

Χαρίτων Ἀφροδισιεύς, Ἀθηναγόρου τοῦ ῥήτορος ὑπογραφεύς, πάθος ἐρωτικὸν ἐν Συρακούσαις γενόμενον διηγήσομαι.

Ἑρμοκράτης ὁ Συρακοσίων στρατηγός, οὗτος ὁ νικήσας Ἀθηναίους, εἶχε θυγατέρα Καλλιρόην τοὔνομα, θαυμαστόν τι χρῆμα παρθένου καὶ ἄγαλμα τῆς ὅλης Σικελίας. ἦν γὰρ τὸ κάλλος οὐκ ἀνθρώπινον ἀλλὰ θεῖον, οὐδὲ Νηρηΐδος ἢ Νύμφης τῶν ὀρειῶν ἀλλ' αὐτῆς Ἀφροδίτης Παρθένου. φή⌈μη δὲ⌉ τοῦ παραδόξου θεάματος πανταχοῦ διέτρεχε καὶ μνηστῆρες κατέρρεον εἰς Συρακούσας, δυνάσται τε καὶ παῖδες τυράννων, οὐκ ἐκ Σικελίας μόνον, ἀλλὰ καὶ ἐξ Ἰταλίας καὶ Ἠπείρου καὶ ἐθνῶν ⌜τῶν⌝ ἐν ἠπείρῳ. ὁ δὲ Ἔρως ⌜ζεῦ⌝γος ἴδιον ἠθέλησε συμπλέξαι. Χαιρέας γάρ τις ἦν μειράκιον εὔ⌜μορφον, πάντων⌝ ὑπερέχον, οἷον Ἀχιλ⌜λέα καὶ Νιρέα καὶ⌝ Ἱππόλυτον καὶ Ἀλκιβιάδην ⌜πλάσται⌝ τε καὶ ⌜γραφεῖς ⟨ἀπο⟩δεικ⌝νύουσι, πατρὸς Ἀρίστωνος τὰ δεύτερα ἐν Συρακούσαις μετὰ Ἑρμοκράτην φερομένου. ⌜καί τις⌝ ἦν ἐν αὐτοῖς πολιτικὸς φθόνος ⌜ὥστε⌝ θᾶττον ἂν πᾶσιν ἢ ἀλλήλοις ἐκήδευσαν. φιλόνεικος δέ ἐστιν ὁ Ἔρως καὶ χαίρει τοῖς παραδόξοις κατορθώμασιν· ἐζήτησε δὲ τοιόνδε τὸν καιρόν.

Ἀφροδίτης ⌜ἑορτὴ δημοτελής, καὶ πᾶσαι σχεδὸν⌝ αἱ γυναῖκες ἀπῆ⌜λθον⌝ εἰς τὸν νεών. τέως δὲ μὴ προϊοῦ⌜σαν τὴν Καλλι⌝ρόην προ⌜ήγαγεν ἡ μήτηρ, ⟨τοῦ πατρὸς⟩ κε⌝λεύσαντος προσκυνῆσαι τὴν θεόν. τότε δὲ Χαιρέας ⌜ἀπὸ τῶν⌝ γυμνασίων ἐβάδιζεν οἴκαδε στίλβων ὥσπερ ἀστήρ· ἐπήνθει γὰρ αὐτοῦ τῷ ⌜λαμπρ⌝ῷ τοῦ προ⌜σώπου τὸ⌝ ἐρύθημα τῆς παλαίστρας ὥσπερ ἀργύρῳ χρυσός. ἐκ τύχης οὖν περί τινα ⌜καμπὴν στενοτέραν⌝ συν⌜αντ⌝ῶ⌜ντες⌝ περιέπεσον ἀλλήλοις, τοῦ θεοῦ πολιτευσαμένου τήνδε τὴν ⌜⟨συνοδίαν⟩ ἵνα ἑκά⟨τερος τῷ⟩ ἑτέρ⟨ῳ⟩⌝ ὀφθῇ. ταχέως οὖν πάθος ἐρωτικὸν ἀντέδωκαν ἀλλή⌜λοις . . . . τοῦ κάλλους ⟨τῇ εὐ⟩γενεί⟨ᾳ⟩ συνελθόντος.

Ὁ μὲν⌝ οὖν Χαιρέας οἴκαδε μετὰ τοῦ τραύματος ⌜μόλις ἀπῄει, καὶ ὥσπερ τις ⟨ἀρισ⟩τεὺς ἐν πολέμῳ τρωθεὶς⌝ καιρίαν, καὶ καταπεσεῖν μὲν αἰδούμενος, στῆναι δὲ μὴ δυν⌜άμεν⌝ος. ⌜ἡ δὲ⌝ παρθένος τῆς Ἀφροδίτης τοῖς ποσὶ προσέπεσε καὶ καταφιλοῦσα, "σύ μοι, δέσπ⌜οινα" εἶπε, "δὸς ἄνδρα τοῦτον ὃν ἔδει⌝ξας." νὺξ ἐπῆλθεν ἀμφοτέροις δεινή· τὸ γὰρ πῦρ ἐξεκαίετο. ⌜δεινότερον δ' ἔπασχ⌝εν ἡ παρθένος διὰ τὴν σιωπήν, αἰδουμένη κατάφωρος γενέσθαι. Χαι⌜ρέας δὲ νεανί⌝ας ⌜εὐφυὴς καὶ μεγαλ⌝όφρων, ἤδη τοῦ σώματος αὐτῷ φθίνοντος, ἀπετόλμησεν εἰπεῖν πρὸς τοὺς γονεῖς ὅτι ἐρᾷ καὶ οὐ βιώσεται τοῦ Καλλιρόης γάμου μὴ τυχών. ἐστέναξεν ὁ πατὴρ ⌜ἀκούσας καὶ "οἴχῃ⌝ δή μοι, τέκνον" ⟨ἔφη⟩· "δῆλον γάρ ἐστιν ὅτι Ἑρμοκράτης οὐκ ἂν δοίη σοὶ τὴν θυγατέρα ⌜τοσούτους ἔχ⌝ων μνηστῆρας ⌜πλουσίους⌝ καὶ βασιλεῖς. οὔκουν οὐδὲ πειρᾶσθαί σε δεῖ, μὴ φανερῶς ὑβρισθῶμεν." εἶθ' ὁ μὲν πατὴρ παρε⌜μυθ⌝εῖτο τὸν παῖδα, τῷ δὲ ηὔξετο τὸ κακὸν ⌜ὥστε⌝ μηδὲ ἐπὶ τὰς συνήθεις προϊέναι διατριβάς. ἐπόθει δὲ τὸ γυμνάσιον Χαιρέαν καὶ ὥσπερ ἔρημον ἦν. ἐφίλει γὰρ αὐτὸν ἡ νεολα⌜ία. πο⌝λυπραγμονοῦντες δὲ τὴν αἰτίαν ἔμαθον τῆς νόσου, καὶ ἔλεος πάντας εἰσῄει μειρακίου καλοῦ κινδυνεύοντος ἀπολέσθαι διὰ πάθος ψυχῆς εὐφυοῦς.

Ἐνέστη νόμιμος ἐκκλησία. συγκαθεσθεὶς οὖν ὁ δῆμος τοῦτο πρῶτον καὶ μ⌈όνον ἐβόα⌉ "καλὸς Ἑρμοκράτης, μέγας στρατηγός, σῶζε Χαιρέαν· τοῦτο πρῶτον τῶν τροπαίων. ἡ πόλις μνηστεύεται τοὺς γάμους σήμερον ἀλλήλων ἀξίων." τίς ἂν μηνύσειε τὴν ἐκκλησίαν ἐκείνην, ἧς ὁ Ἔρως ἦν δημαγωγός; ἀνὴρ δὲ φιλόπατρις Ἑρμοκράτης ἀντειπεῖν οὐκ ἠδυνήθη τῇ πόλει δεομένῃ. κατανεύσαντος δὲ αὐτοῦ πᾶς ὁ δῆμος ἐξεπήδησε τοῦ θεάτρου, καὶ οἱ μὲν νέοι ἀπῄεσαν ἐπὶ Χαιρέαν, ἡ βουλὴ δὲ καὶ οἱ ἄρχοντες ἠκολούθουν Ἑρμοκράτει· παρῆσαν δὲ καὶ αἱ γυναῖκες αἱ Συρακοσίων ἐπὶ τὴν οἰκίαν νυμφαγωγοῦσαι. ὑμέναιος ᾔδετο κατὰ πᾶσαν τὴν πόλιν· μεσταὶ δὲ αἱ ῥῦμαι στεφάνων, λαμπάδων· ἐρραίνετο τὰ πρόθυρα οἴνῳ καὶ μύροις. ἥδιον ταύτην ⟨τὴν⟩ ἡμέραν ἤγαγον οἱ Συρακόσιοι τῆς τῶν ἐπινικίων. ἡ δὲ παρθένος οὐδὲν εἰδυῖα τούτων ἔρριπτο ἐπὶ τῆς κοίτης ἐγκεκαλυμμένη, κλαίουσα καὶ σιωπῶσα. προσελθοῦσα δὲ ἡ τροφὸς τῇ κλίνῃ "τέκνον" εἶπε, "διανίστασο, πάρεστι γὰρ ἡ εὐκταιοτάτη πᾶσιν ἡμῖν ἡμέρα· ἡ πόλις σε νυμφαγωγεῖ."

Τῆς δ' αὐτοῦ λύτο γούνατα καὶ φίλον ἦτορ·

οὐ γὰρ ᾔδει, τίνι γα⌈μεῖτ⌉αι. ἄφωνος εὐθὺς ἦν καὶ σκότος αὐτῆς τῶν ὀφθαλμῶν κατεχύθη καὶ ὀλίγου δεῖν ἐξέπνευσεν· ἐδόκει δὲ τοῦτο τοῖς ὁρῶσιν αἰδώς. ἐπεὶ δὲ ταχέως ἐκόσμησαν αὐτὴν αἱ θεραπαινίδες, τὸ πλῆθος ἐπὶ τῶν θυρῶν ἀπέλιπον· οἱ δὲ γονεῖς τὸν νυμφίον εἰσήγαγον πρὸς τὴν παρθένον. ὁ μὲν οὖν Χαιρέας προσδραμὼν αὐτὴν κατεφίλει, Καλλιρόη δὲ γνωρίσασα τὸν ἐρώμενον, ὥσπερ τι λύχνου φῶς ἤδη σβεννύμενον ἐπιχυθέντος ἐλαίου πάλιν ἀνέλαμψε καὶ μείζων ἐγένετο καὶ κρείττων. ἐπεὶ δὲ προῆλθεν εἰς τὸ δημόσιον, θάμβος ὅλον τὸ πλῆθος κατέλαβεν, ὥσπερ Ἀρτέμιδος ἐν ἐρημίᾳ κυνηγέταις ἐπιστάσης· πολλοὶ δὲ τῶν παρόντων καὶ προσεκύνησαν. πάντες δὲ Καλλιρόην μὲν ἐθαύμαζον, Χαιρέαν δὲ ἐμακάριζον.

[My name is Chariton, of Aphrodisias, and I am clerk to the attorney Athenagoras. I am going to tell you the story of a love affair that took place in Syracuse.

The Syracusan general Hermocrates, the man who defeated the Athenians, had a daughter called Callirhoe. She was a wonderful girl, the pride of all Sicily; her beauty was more than human, it was divine, and it was not the beauty of a Nereid or mountain nymph at that, but of the maiden Aphrodite herself. Report of the astonishing vision spread everywhere, and suitors flocked to Syracuse, rulers and tyrants' sons, not just from Sicily but from southern Italy too and farther north, and from foreigners in those parts. But Eros intended to make a match of his own devising. There was a young man called Chaereas, surpassingly handsome, like Achilles and Nireus and Hippolytus and Alcibiades as sculptors and painters portray them. His father, Ariston, was second only to Hermocrates in Syracuse, and the two were political rivals, so that they would have made a marriage alliance with anyone rather than with each other. But Eros likes to win and enjoys succeeding against the odds. He looked for his opportunity and found it as follows.

A public festival of Aphrodite took place, and almost all the women went to her temple. Callirhoe had never been out in public before, but her father wanted her to do reverence to the goddess, and her mother took her. Just at that time Chaereas was walking home from the gymnasium; he was radiant as a star, the flush of exercise blooming on his bright countenance like gold on silver. Now, chance would have it that at the corner of a narrow street the two walked straight into each other; the god had contrived the meeting so that each could see the other. At once they were both smitten with love . . . beauty had met nobility.

Chaereas, so stricken, could barely make his way home; he was like a hero mortally wounded in battle, too proud to fall but too weak to stand. The girl, for her part, fell at Aphrodite's feet and kissed them. "Mistress," she cried, "give me the man you showed me for my husband!" When night came, it brought suffering to both, for the fire was raging in them. The girl suffered more, bcause she could not bear to give herself away and so said nothing to anyone. But when Chaereas began to waste away bodily, he found courage, as befitted a youth of noble and generous disposition, to tell his parents that he was in love and would die if he did not marry Callirhoe. At this his father groaned and said: "Then I have lost you, my boy! Hermocrates would certainly never give you his daughter when he has so many rich and royal suitors for her. You must not even try to win her, or we shall be publicly insulted." Then the father tried to comfort his son, but his illness grew so serious that he did not even go out and follow his usual pursuits. The gymasium missed Chaereas; it was almost deserted, for he was the idol of the young folk. They asked after him, and when they found out what had made him ill, they all felt pity for a handsome youth who looked as if he would die because his noble heart was broken.

A regular assembly took place at this time. When the people had taken their seats, their first and only cry was: "Noble Hermocrates, great general, save Chaereas! That will be your finest monument! The city pleads for the marriage, today, of a pair worthy of each other!" Who could describe that assembly? It was dominated by Eros. Hermocrates loved his country and could not refuse what it asked. When he gave his consent, the whole meeting rushed from the theater; the young men went off to find Chaereas, the council and archons escorted Hermocrates, and the Syracusans' wives too went to his house, to attend the bride. The sound of the marriage hymn pervaded the city, the streets were filled with garlands and torches, porches were wet with wine and perfume. The Syracusans celebrated this day even more joyously than the day of their victory.

The girl knew nothing of all this; she lay on her bed, her face covered, crying and uttering not a word. Her nurse came to her as she lay there. "Get up, my child," she said. "The day we have all been praying so hard for has come: the city is here to see you married!"

> And then her limbs gave way, her heart felt faint,

for she did not know whom she was going to marry. She fainted there and then; darkness veiled her eyes, and she almost expired; the spectators thought it maidenly modesty. As soon as her maids had dressed her, the crowd at the door went away, and his parents brought the bridegroom in to the girl. Well, Chaereas ran to her and kissed her; and when she saw it was the man she loved, Callirhoe, like the flame in a lamp that is on the point of going out and has oil poured into it, at once grew bright again and bigger and stronger. When she appeared in public the whole crowd was struck with wonder, as when Artemis appears to hunters in lonely places; many of those present actually went down on their knees in worship. They all thought Callirhoe beautiful and Chaereas lucky.]

Abbreviations

ANRW	Aufstieg und Niedergang der Römanischen Welt
ATR	Anglican Theological Review
CBQ	Catholic Biblical Quarterly
CJ	Classical Journal
GRBS	Greek, Roman, and Byzantine Studies
HTR	Harvard Theological Review
JBL	Journal of Biblical Literature
JHS	Journal of Hellenic Studies
JSNT	Journal for the Study of the New Testament
NLH	New Literary History
NTS	New Testament Studies
SBLDS	Society of Biblical Literature Dissertation Studies
SPCK	Society for Promoting Christian Knowledge
STR	Sewanee Theological Review
WUNT	Wissenschaftliche Untersuchungen zum Neuen Testament

Bibliography

PRIMARY SOURCES

The usual texts and translations of Holy Scripture and the rabbinic writings are assumed.

Adam and Eve, Life of. Translated by L. S. A. Wells. Revised by M. Whittaker. In H. D. F. Sparks, ed., *The Old Testament Apocrypha*, 147–61. Oxford: Clarendon, 1984. Also translated by M. D. Johnson. In James H. Charlesworth, ed., *The Old Testament Pseudepigrapha*, vol. 2, 379–99. New York: Doubleday, 1985.

Andrewes, Lancelot. *Ninety-Six Sermons*. 5 vols. Oxford: John Henry Parker, 1841.

Antiphanes, *Sappho*. Text and translation in John Maxwell Edmonds, ed., *The Fragments of Attic Comedy*. 3 vols. Leiden: E. J. Brill, 1959. 2.262–65.

Apocalypse of Moses. Translated by M. D. Johnson. In James H. Charlesworth, ed., *The Old Testament Pseudepigrapha*, vol. 2, 379–99. New York: Doubleday, 1985.

Aristophanes, *Knights*. Text and translation in Alan H. Sommerstein, ed., and trans., *The Comedies of Aristophanes*, vol. 2. *Knights*. Warminster: Aris and Phillips, 1981.

Aristotle, *Poetics*. Text in D. W. Lucas, ed., *Aristotle: Poetics*. Oxford: Oxford University Press, 1959. Text and translation in Ingram Bywater, *Aristotle: On the Art of Poetry*. Oxford: Clarendon, 1909.

Barnabas, Epistle of. Text and translation in Kirsopp Lake, *The Apostolic Fathers*. Loeb Classical Library. Cambridge, Harvard University Press; London: Heinemann, 1959.

2 Baruch, (*Syriac Apocalypse of*). Translation by A. F. J. Klijn. In James H. Charlesworth, ed., *The Old Testament Pseudepigrapha*, vol. 1, 615–52. New York: Doubleday, 1983.

The Battle of Maldon. Text in Elliot Van Kirk Dobbie, ed., *The Anglo-Saxon Minor Poems*, vol. 6. New York: Columbia University Press, 1942, 7–16; Translation in Kevin Crossley-Holland, *The Battle of Maldon and Other Old English Poems.* London: Macmillan, 1965. 29–38.

Caedmon's Hymn. Text in Elliot Van Kirk Dobbie, ed., *The Anglo-Saxon Minor Poems.* New York: Columbia University Press, 1942, 105–106; Translation in Kevin Crossley-Holland, *The Battle of Maldon and Other Old English Poems.* London: Macmillan, 1965. 95.

Chariton. *Callirhoe.* Text in Warren E. Blake, ed., *Charitonis Aphrodisiensis: De Chaerea et Callirhoe Amatoriarum Narrationum Libri Octo.* Oxford: Clarendon, 1938. Translated by B. P. Reardon. In B. P. Reardon, ed., *Collected Ancient Greek Novels.* Berkeley and Los Angeles, University of California Press, 1989.

1 Clement. Text in A. Jaubert, ed., *Clément de Rome. Epitre aux Corinthiens* [*Sources chrêtiennes* 167]. Paris: Cerf, 1971. Text and translation in Kirsopp Lake, *The Apostolic Fathers.* Loeb Classical Library. Cambridge, Harvard University Press; London: Heinemann, 1959.

2 Clement. Text in K. Bihlmeyer and W. Schneemelcher (post F. X. Funk), *Die Apostolischen Väter.* 3d ed. Tübingen: Mohr, 1970. Text and translation in Kirsopp Lake, *The Apostolic Fathers.* Loeb Classical Library. Cambridge: Harvard University Press; London: Heinemann, 1959.

"Demetrius." *On Style.* Text in Ludovic Radermacher, *Demetrii Phalerei qui dicitur de Elocutione Libellus.* Stuttgart: Teubner, 1967 (reprint of 1901 ed.). Text and translation in W. Hamilton Fyfe and W. Rhys Roberts, *Aristotle: the Poetics; "Longinus" On the Sublime; Demetrius on Style.* London: Heinemann; New York: Putnam, 1932. Translation in G. M. A. Grube, *A Greek Critic: Demetrius on Style.* Toronto, University of Toronto Press, 1961.

Eusebius. *Ecclesiastical History.* Text in Gustave Bardy, ed., *Eusèbe de Césarée. Histoire Ecclésiastique.* 3 vols. [*Sources chrétiennes* 31, 41, 55]. Paris: Cerf, 1952–58 (3d reprint, 1967). Text and translation in Kirsopp Lake and J. E. L. Oulton, *Eusebius: Ecclesiastical History.* 2 vols. New York: Putnam; London: Heinemann, 1926–32.

Ad Herennium. Text and translation in Harry Caplan, [*Cicero*]: *Ad Herennium de Ratione Dicendi* (*Rhetorica ad Herennium*). Loeb Classical Library. Cambridge: Harvard University Press; London: Heinemann, 1989.

Homer. *Iliad.* Text in Thomas W. Allen, ed., *Homeri Ilias.* Oxford: Clarendon, 1931. Text and translation in A. T. Murray, *Homer: The Iliad.* 2 vols. Loeb Classical Library. Cambridge: Harvard University Press; London: Heinemann, 1944.

———. *Odyssey.* Text in Thomas W. Allen, ed., *Odyssey.* Oxford Classical Texts. Oxford: Clarendon, 1906. Text and translation in A. T. Murray,

Homer: The Odyssey. 2 vols. Loeb Classical Library. Cambridge: Harvard University Press; London: Heinemann, 1938.

Horace. *The Art of Poetry.* Text in C. O. Brink, *Horace on Poetry.* 3 vols. Cambridge: Cambridge University Press, 1963–82. Text and translation in H. Rushton Fairclough, *Horace: Satires, Epistles, and Ars Poetica.* Loeb Classical Library. Cambridge: Harvard University Press; London: Heinemann, 1939.

Joseph and Asenath. Translated by C. Burchard. In James H. Charlesworth, ed., *The Old Testament Pseudepigrapha,* vol. 2, 177–247. New York: Doubleday, 1985.

Josephus. *The Jewish War and Jewish Antiquities.* Text in Benedikt Niese, ed., *Flavii Iosephi opera.* 6 vols. Berlin: Weidmann, 1887–95 (reprint, vols. 4–6, 1955). Text and translation in H. St. J. Thackeray, Ralph Marcus, Louis H. Feldman, *Josephus.* Loeb Classical Library. 10 vols. Cambridge: Harvard University Press; London: Heinemann, 1926–65.

Justin Martyr. *Apology, Second Apology,* and *Dialogue with Trypho.* Text in Edgar J. Goodspeed, *Die ältesten Apologeten.* Göttingen: Vandehoeck and Ruprecht, 1915. For text of the Apologies, see also A. W. F. Blunt, *The Apologies of Justin Martyr.* Cambridge: Cambridge University Press, 1911. Translation by Marcus Dods in Alexander Roberts and James Donaldson, eds., *Ante-Nicene Christian Library.* Edinburgh: T and T. Clark, 1867.

Kerygma Petrou. Text in Ernst von Dobschütz, *Das Kerygma Petrou kritisch untersucht. Texte und Untersuchungen zur Geschichte der altchristlichen Literatur* 11.1. Leipzig: Hinrichs'sche Buchhandlung, 1893, 18–27. Translation in Edgar Hennecke and Wilhelm Schneemelcher, *New Testament Apocrypha.* 2 vols. London: Lutterworth; Philadelphia: Westminster, 1963 (vol. 1), 1965 (vol. 2).

[Longinus]. *On the Sublime.* Text in Donald A. Russell, *"Longinus." On the Sublime.* Oxford: Clarendon, 1964. Text and translation in W. Hamilton Fyfe and W. Rhys Roberts, *Aristotle: The Poetics; "Longinus" On the Sublime; Demetrius on Style.* London: Heinemann; New York: Putnam, 1932.

Lucian. *Demonax, The Passing of Peregrinus,* and *How to Write History.* Text in M. D. McLeod, ed., *Luciani Opera.* 4 vols. Oxford: Oxford University Press, 1972–87. Text and translation in A. M. Harmon, *Lucian.* 8 vols. Loeb Classical Library. Cambridge: Harvard University Press; London: Heinemann, 1913.

Lucretius. *De Rerum Natura.* Text and translation in W. H. D. Rouse, *Lucretius: De Rerum Natura.* Loeb Classical Library. Cambridge: Harvard University Press; London: Heinemann, 1937.

Malory, Sir Thomas. *Le Morte Darthur*. Text in Eugène Vinaver, *The Works of Sir Thomas Malory*. 3 vols. Oxford: Clarendon, 1948 (corrected reprint). Normalized spelling (and Caxton's arrangement by book and chapter) in Ernest Rhys, ed., *Le Morte D'Arthur by Sir Thomas Malory*. 2 vols. Everyman's Library. London and Toronto: Dent; New York: Dutton, 1906.

Nepos, *Cato* and *Atticus*. Text and translation in J. C. Rolfe, *Cornelius Nepos*. Loeb Classical Library. Cambridge: Harvard University Press; London: Heinemann, 1984. English translation in Nicholas Horsfall, *Cornelius Nepos: A Selection, Including the Lives of Cato and Atticus*. Oxford: Clarendon, 1989.

Peter, Gospel of. Text in M. G. Mara, ed., *Evangile de Pierre* [*Sources Chrétiennes* 201]. Paris: Cerf, 1973. Translation in Edgar Hennecke and Wilhelm Schneemelcher, *New Testament Apocrypha*. 2 vols. London: Lutterworth; Philadelphia: Westminster, 1963.

Philo. *On the Life of Moses*. Text in Leopold Cohn and Paul Wendland, (eds.), *Philonis Alexandrini opera quae supersunt*. 6 vols. Berlin : Reimar, 1896 (reprint De Gruyter, 1962). Text and translation in F. H. Colson, *Philo*. 6 vols. Loeb Classical Library. Cambridge: Harvard University Press; London: Heinemann, 1935.

Philostratus. *Life of Apollonius of Tyana*. Text in Carl Ludwig Kayser, ed., *Flavii Philostrati opera*. Vol. 1. Leipzig: Teubner, 1870 (reprint Hildersheim: Olms, 1964). Text and translation in F. C. Conybeare, *Philostratus: The Life of Apollonius of Tyana*. 2 vols. London: Macmillan, 1912.

Pindar, Life of. Text in Anders B. Drachmann, ed., *Scholia Vetera in Pindari Carmina* 1. Amsterdam: Hakkert, 1969 (reprint) 1–3. Translation in Mary R. Lefkowitz, *The Lives of the Greek Poets*, 155–57. Baltimore, Md.: John Hopkins University Press, 1981.

Pliny the Younger. *Letters*. Text and translation in Betty Radice, *Pliny: Letters and Panegyricus*. 2 vols. Loeb Classical Library. Cambridge: Harvard University Press; London: Heinemann, 1969.

Plutarch. *Parallel Lives*. Text in K. Ziegler, ed., *Plutarchi vitae parallelae*. 3 vols. 4th ed. (incomplete). Leipzig: Teubner, 1969. Texts of remaining "Lives," and text and translation, in Bernadotte Perrin, *Plutarch's Lives*. 11 vols. London: Heinemann; New York: Putnam, 1919.

Prophets, The Lives of the. Translation by D. R. A. Hare. In James H. Charlesworth, ed., *The Old Testament Pseudepigrapha*, vol 2, 379–99. New York: Doubleday, 1985.

Pseudo-Philo. *Biblical Antiquities*. Text edited by Daniel J. Harrington and French translation by Jacques Caseaux (reviewed by Charles Perrot and Pierre-Maurice Bogaert), *Pseudo-Philon: Les Antiquités Bibliques*.

Vol. 1. [*Sources chrétiennes*]. Paris: Cerf, 1976. English translation by D. J. Harrington. In James H. Charlesworth, ed., *The Old Testament Pseudepigrapha*, Vol. 2, 297–377. New York: Doubleday, 1985.

Quintillian. *Institutio Oratoria*. Text and translation in H. E. Butler, *The Institutio Oratoria of Quintillian*. 4 vols. London: Heinemann, 1921–33.

Secundus the Philosopher, Life of. Text and translation in Ben Edwin Perry, *Secundus the Silent Philosopher*. APA Monographs 22. Ithaca: American Philological Association and Cornell University Press. Translation in David Aune, ed., *Greco-Roman Literature and the New Testament*. Sources for Biblical Study 21. Atlanta, Ga.: Scholars, 1988.

Seneca. *To Lucilius*. Text and translation in R. M. Gummere, *Seneca: Epistulae Morales*. 3 vols. Cambridge: Harvard University Press, 1917, 1920, 1925.

Suetonius. *Lives of the Caesars*. Text and translation in J. C. Rolfe, *Suetonius*. 2 vols. London: Heinemann; New York: Macmillan, 1914.

Tacitus. *Agricola*. Text in R. M. Ogilvie and Ian Richmond, *Cornelii Taciti: De Vita Agricolae*. Oxford: Clarendon, 1967. Text and translation in William Peterson, *Tacitus: Dialogus, Agricola, Germania*. London: Heinemann; New York: Putnam, 1932.

———. *The Annals*. Text and translation in Clifford H. Moore and John Jackson, *Tacitus: The Histories, The Annals*. 4 vols. Loeb Classical Library. Cambridge: Harvard University Press; London: Heinemann, 1937.

———. *Dialogue on Oratory*. Text and translation in William Peterson, *Tacitus: Dialogus, Agricola, Germania*. London: Heinemann; New York: Putnam, 1932.

Testament of Isaac. Translation by W. F. Winespring. In James H. Charlesworth, ed., *The Old Testament Pseudepigrapha*, vol. 1, 903–11. New York: Doubleday, 1983.

Testaments of the Twelve Patriarchs. Translation by H. C. Kee. In James H. Charlesworth, ed., *The Old Testament Pseudepigrapha*, vol. 1, 775–828. New York: Doubleday, 1983.

Theon. *Progymnasmata: Peri Chreias*. Text in Leonard Spengel, *Rhetores Graeci*, 59–130. Leipzig: Teubner, 1854. Text and translation in Ronald F. Hock and Edward N. O'Neil, *The Chreia in Ancient Rhetoric*, Vol. 1, *The Progymnasmata*. Atlanta, Ga.: Scholars, 1986.

Thomas, Gospel of. Text and translation in Bentley Layton, ed., *Nag Hammadi Codex II, 2–7*. Critical text by Bentley Layton, translation by Thomas O. Lambdin, introduction by Helmut Koester. Leiden: Brill, 1989. Also John S. Kloppenborg, Marvin W. Myer, Stephen J. Patterson, and Michael G. Steinhauser, *Q-Thomas Reader*. Sonoma, Calif.: Polebridge, 1990.

Xenophon of Ephesus. *An Ephesian Tale*. Text in Antonius D. Papanikolaou, *Xenophontis Ephesii: Ephesiacorum Libri V de Amoribus Anthiae et Abrocomae*. Leipzig: Teubner, 1973. English translation by Graham Anderson. In B. P. Reardon, ed., *Collected Ancient Greek Novels*. Berkeley and Los Angeles, University of California Press, 1989.

SECONDARY SOURCES

Achtemeier, Paul J. *Mark*. Philadelphia: Fortress, 1986.

———. "*Omne verbum sonat:* The New Testament and the Oral Environment of Late Western Antiquity." *JBL* 109.1 (1990): 3–27.

Alexander, Philip S. "Midrash and the Gospels." In C. M. Tuckett, ed., *Synoptic Studies: The Ampleforth Conferences of 1982 and 1983,* 1–18. JSNT Supplement Series 7. Sheffield: JSOT, 1984.

———. "Rabbinic Biography and the Biography of Jesus: A Survey of the Evidence." In C. M. Tuckett, ed., *Synoptic Studies: The Ampleforth Conferences of 1982 and 1983,* 19–50. JSNT Supplement Series 7. Sheffield: JSOT, 1984.

Alter, Robert. *The Art of Biblical Narrative*. New York: Basic Books, 1981.

Auerbach, Eric. *Mimesis: The Representation of Reality in Western Literature*. Translated by Willard R. Trask. Princeton: Princeton University Press, 1953.

Aune, David A. *The New Testament in Its Literary Environment*. Philadelphia: Westminster, 1987.

———. "The Problem of the Genre of the Gospels: A Critique of C. H. Talbert's *What is a Gospel?*" In R. T. France and D. Wenham, eds., *Gospel Perspectives: Studies of History and Tradition in the Four Gospels* 2, Sheffield: JSOT, 1981. pp. 9–60.

———, ed., *Greco-Roman Literature and the New Testament*. Atlanta, Ga.: Scholars, 1988.

Barth, Karl. *Dogmatics in Outline*. Translated by G. T. Thompson. London: SCM, 1949.

Bartholomew, Gilbert L.: *see* Boomershine, Thomas E.

Beavis, Mary Ann. *Mark's Audience: The Literary and Social Setting of Mark 4:11–12*. Sheffield: Sheffield Academic, 1989.

Beck, Norman A. "Reclaiming a Bibical Text: The Mark 8.14–21 Discussion about Bread in the Boat." *CBQ* 43 (1981): 49–56.

Benson, Larry. "The Literary Character of Anglo-Saxon Formulaic Poetry." *PMLA* 81 (1966): 334–41.

Bilezikian, Gilbert G. *The Liberated Gospel: A Comparison of the Gospel of Mark and Greek Tragedy*. Grand Rapids, Mich.: Baker, 1977.

Black, C. Clifton. *The Disciples According to Mark: Markan Redaction in Current Debate*. JSNT Supplement Series 27. Sheffield: JSOT, 1989.

Bonner, Stanley F. *Education in Ancient Rome: From the Elder Cato to the Younger Pliny*. Berkeley and Los Angeles: University of California Press, 1977.

Boomershine, Thomas E. "Mark 16:8 and the Apostolic Commission." *JBL* 100.2 (1981): 225–39.

———. "Peter's Denial as Polemic or Confession: The Implications of Media Criticism for Biblical Hermeneutics." In Lou H. Silberman, ed., *Orality, Aurality, and Biblical Narrative,* 47–68. Semeia 39. Decatur, Ga.: Scholars, 1987.

———, and Gilbert L. Bartholomew. "The Narrative Technique of Mark 16:8." *JBL* 100.2 (1981): 213–23.

Boring, M. Eugene. "Mark 1:1–15 and the Beginning of the Gospel." In Dennis E. Smith, ed., *How Gospels Begin*, 43–81. Semeia 52. Atlanta, Ga.: Scholars, 1991.

Botsford, Keith. "Saul Bellow: Made in America." *The Independent* (London). February 10, 1990, p. 29.

Bozarth-Campbell, Alla. *The Word's Body: An Incarnational Aesthetic of Interpretation*. Tuscaloosa: University of Alabama Press, 1979.

Branick, Vincent. *The House Church in the Writings of Paul*. Wilmington, Del.: Glazier, 1989.

Brown, Raymond E. "Jesus and Elijah." *Perspective* 12 (1971): 85–104.

———, and John P. Meier. *Antioch and Rome: New Testament Cradles of Catholic Christianity*. New York and Ramsey, New Jersey: Paulist, 1983.

Bryan, Christopher. "The Judicious Mr. Hooker and the Early Christians: The Relationship of Scripture and Reason in the First Century of the Christian Era." In Donald W. Armentrout, ed., *This Sacred History: Anglican Reflections for John Booty*, 144–60. Cambridge, Mass.: Cowley, 1990.

———. "The Preachers and the Critics: Reflections on Historical Criticism." *ATR* 74.1 (1992): 37–53.

Bultmann, Rudolf. *History of the Synoptic Tradition*. New York: Harper and Row, 1963.

Burridge, Richard A. "A Genre for the Gospels: The Biographical Character of Matthew." (Review). *Anvil* 2.2 (1985): 179–80.

———. *What Are the Gospels? A Comparison with Graeco-Roman Biography*. SNTS Monograph Series 70. Cambridge: Cambridge University Press, 1992.

Carrington, Philip. *The Primitive Christian Calendar: A Study in the Making of the Markan Gospel.* Cambridge: Cambridge University Press, 1952.

Chaytor, H. J. *From Script to Print: An Introduction to Medieval Vernacular Literature.* New York: October House, 1967.

Chilton, Bruce. *The Glory of Israel.* Sheffield: JSOT, 1983.

———, and J. I. H. McDonald. *Jesus and the Ethics of the Kingdom.* London: SPCK, 1987.

Clarke, M. L. *Higher Education in the Ancient World.* London: Routledge and Kegan Paul, 1971.

Collins, Adela Yarbro. "The Composition of the Passion Narrative in Mark" *STR* 36.1 (1992): 57–77.

Collins, John N. *Diakonia: Re-interpreting the Ancient Sources.* New York: Oxford University Press, 1990.

Cox, Patricia. *Biography in Late Antiquity: A Quest for the Holy Man.* Berkeley and Los Angeles: University of California Press, 1983.

Cranfield, C. E. B. *The Gospel According to Saint Mark.* Cambridge Greek New Testament Commentary. 3d ed. with supplementary notes. Cambridge: Cambridge University Press, 1966.

Crossan, John Dominic. *In Parables: The Challenge of the Historical Jesus.* New York: Harper and Row, 1973.

Crossan, John R., S.J. *The Gospel in Parable: Metaphor, Narrative and Theology in the Synoptic Gospels.* Philadelphia: Fortress, 1988.

Daube, David. *The New Testament and Rabbinic Judaism.* London: Athlone, 1956.

Davidson, O. M. "A Formulaic Analysis of Samples Taken from the *Shâhnâma* of Firdowsi." *Oral Tradition* 3.1–2 (1988): 88–105.

Davies, W. D. "Reflections on Archbishop Carrington's *The Primitive Christian Calendar.*" In W. D. Davies and David Daube, eds., *The Background of the New Testament and Its Eschatology. Studies in Honour of C. H. Dodd.* Cambridge: Cambridge University Press, 1956, 124–52.

Davis, Lindsey. *Silver Pigs.* New York: Ballentine, 1991.

Denniston, J. D. *Greek Literary Criticism.* London: Dent; New York: Dutton, 1924.

Dewey, Joanna. *Markan Public Debate: Literary Technique, Concentric Structure, and Theology in Mark 2:1–3:6.* SLBDS 48. Chico, Calif.: Scholars, 1980.

———. "Mark as Interwoven Tapestry: Forecasts and Echoes for a Listening Audience." *CBQ* 53.2 (1991): 221–36.

———. "Oral Methods of Structuring Narrative in Mark" *Interpretation* 43.1 (1989): 32–44.

Dodd, Charles H. "The Framework of the Gospel Narrative." *Expository Times*

43 (1932) 396–400, reprinted in C. H. Dodd, *New Testament Studies*, 1–11. Manchester: Manchester University Press, 1953.

———. *The Parables of the Kingdom*. London: Nisbet, 1936.

Donahue, John R., S.J. *The Gospel in Parable: Metaphor, Narrative, and Theology in the Synoptic* Gospels. Philadelphia: Fortress, 1988.

Dowd, Sharyn Echols. *Prayer, Power, and the Problem of Suffering: Mark 11:22–25 in the Context of Markan Theology*. Atlanta, Ga.: Scholars, 1988.

Downing, F. Gerald. "A bas Les Aristos. The Relevance of Higher Literature for the Understanding of the Earliest Christian Writings." *Novum Testamentum* 30.3 (1988): 212–30.

———. *Christ and the Cynics: Jesus and Other Radical Preachers in First-Century Tradition*. Manuals 4. Sheffield: JSOT, 1988.

Droge, Arthur J. "Call Stories in Greek Biography and the Gospels." In Kent Harold Richards, ed., SBL 1983 Seminar Papers, Scholars Press: Chicago, 1983, 245–57.

Duff, Paul Brooks. "The March of the Divine Warrior and the Advent of the Greco-Roman King: Mark's Account of Jesus' Entry into Jerusalem." *JBL* 3.1 (1992) 55–71.

Easterling, P. E., and Bernard M. W. Knox. *Cambridge History of Classical Literature*. Vol. 1. London: Cambridge University Press, 1985.

Farrell, Thomas J., S.J., "Kelber's Breakthrough." In Lou H. Silberman, ed., *Orality, Aurality, and Biblical Narrative*, 27–46. Semeia 39. Decatur, Ga.: Scholars, 1987.

Fowler, Alastair. *Kinds of Literature: An Introduction to the Theory of Genres and Modes*. Oxford: Oxford University Press, 1982.

———. "The Life and Death of Literary Forms." In Ralph Cohen, ed., *New Directions in Literary History*, 77–105. London: Routledge and Kegan Paul; Baltimore, Md.: Johns Hopkins University Press, 1974.

Fowler, Robert M. *Let the Reader Understand: Reader-Response Criticism and the Gospel of Mark*. Minneapolis, Minn.: Fortress, 1991.

———. "Loaves and Fishes: The Function of the Feeding Stories in the Gospel of Mark." SBLDS 54. Chico, Calif.: Scholars, 1981.

Frei, Hans. *The Identity of Jesus Christ: An Enquiry into the Hermeneutical Basis of Dogma*. Philadelphia: Fortress, 1974.

Fry, Donald K. "Caedmon as a Formulaic Poet." In *Forum for Modern Language Studies* 10 (1974): Reprinted in Joseph J. Duggan, ed., *Oral Literature: Seven Essays*, 41–46. Edinburgh: Scottish Academic Press; New York: Barnes and Noble, 1975.

———. "The Memory of Caedmon." In John Miles Foley, ed., *Oral Traditional Literature: A Festschrift for Albert Bates Lord*, 282–93. Columbus, Ohio: Slavica, reprint 1983.

Fuller, Reginald H. *The Formation of the Resurrection Narratives*. New York: Macmillan; London: Collier-Macmillan, 1971.

———. *He that Cometh: The Birth of Jesus in the New Testament*. Harrisburg, Pa.: Morehouse, 1990.

Funk, Robert W., (ed.) *New Gospel Parallels*. Vol. 1. *The Synoptic Gospels*. Philadelphia: Fortress, 1985.

Georgi, Dieter. *The Opponents of Paul in 2 Corinthians*. Philadelphia: Fortress, 1986.

Gill, Christopher. "The Character-Personality Distinction." In Christopher Pelling, ed., *Characterization and Individuality in Greek Literature*, 1–31. Oxford: Clarendon, 1990.

Globe, Alexander. "The Caesarean Omission of the Phrase 'Son of God' in Mark 1.1." *HTR* 75.2 (1982): 209–18.

Goulder, Matthew. *Midrash and Lection in Matthew*. London: SPCK, 1974.

Green, Joel B. *The Death of Jesus: Tradition and Interpretation in the Passion Narrative*. WUNT 2/33; Tübingen: Mohr (Siebeck), 1988.

Grimes, J. E. *The Thread of Discourse*. The Hague: Mouton, 1975.

Guelich, Robert. "The Gospel Genre." In Peter Stuhlmacher, ed., *Das Evangelium und die Evangelien: Vorträge vom Tübinger Symposium 1982*, 183–219. WUNT 28. Tübingen: Mohr, 1983.

Haenchen, Ernst. *Der Weg Jesu*. Berlin: Töpelmann, 1966.

Hägg, Tomas. *The Novel in Antiquity*. Oxford: Basil Blackwell; Berkeley and Los Angeles: University of California Press, 1983.

Hare, Douglas R. A. *The Son of Man Tradition*. Minneapolis: Fortress, 1990.

Hartmann, Geoffrey H., and Sanford Budick, eds., *Midrash and Literature*. New Haven: Yale University Press, 1986.

Havelock, Eric A. "Oral Composition in the *Oedipus Tyrranus* of Sophocles." *NLH* 16(1984): 175–97.

———. *Preface to Plato*. Cambridge: Harvard University Press, Belknap, 1963.

Head, Peter M. "A Text-Critical Study of Mark 1.1 'The Beginning of the Gospel of Jesus Christ'." *NTS* 37.4 (1991): 621–29.

Heil, John Paul. "Reader-Response and the Narrative Context of the Parables about Growing Seed in Mark 4:1–34." *CBQ* 54.2 (1992) 271–86.

Hendrickson, G. L. "Ancient Reading." *CJ* 25 (1929): 182–96.

Hengel, Martin. *Studies in the Gospel of Mark*. Philadelphia: Fortress, 1985.

———. *The Zealots: Investigations into the Jewish Freedom Movement in the Period from Herod I until 70 A.D.* Edinburgh: T & T Clark, 1989. Translation of *Die Zeloten: Untersuchungen zur Jüdischen Freiheitsbewegung in der Zeit von Herodes I. 70 n. Chr.* Leiden: Brill 1961; 2d ed., rev., 1976.

Hennecke, Edgar, and Wilhelm Schneemelcher. *New Testament Apocrypha*. 2 vols. London: Lutterworth; Philadelphia: Westminster, 1963.

Hirsch, E. D. *Validity in Interpretation*. New Haven: Yale University Press, 1967.

Hooker, Morna D. *The Son of Man in Mark: A Study of the Background of the Term "Son of Man" and Its Use in St. Mark's Gospel*. London: SPCK, 1967.

———. *A Commentary on the Gospel according to St. Mark*. London: A. and C. Black, 1991.

Horsely, Richard A., and John S. Hanson. *Bandits, Prophets, and Messiahs: Popular Movements in the Time of Jesus*. Minneapolis, Minn.: Winston, 1985.

Hurtado, Larry W. *Text-Critical Methodology and the Pre-Caesarean Text: Codex W in the Gospel of Mark*. Grand Rapids, Mich.: Eerdmans, 1981.

Iersel, B. van. "De betekenis van Marcus vanuit zijn topografische structuur." *Tijdschrift voor Theologie* 22 (1982): 117–38.

Jeremias, Joachim. *The Parables of Jesus*. London: SCM, 1976. (1st German ed., Zurich: Zwingli, 1947.)

Jonge, Marinus de. "Messianic Ideas in Later Judaism." Parts 1, 2, 3 and 5 in Gerhard Friedrich, *Theological Dictionary of the New Testament,* vol. 9, 509–17, 520–21. Grand Rapids, Mich.: Eerdmans, 1974.

Juel, Donald. *Messiah and Temple: The Trial of Jesus in the Gospel of Mark*. SBL Dissertation Series. Missoula, Mont.: Scholars, 1977.

Kähler, Martin. *Der Sogenannte historische Jesus und der geschichtliche biblische Christus.* 2d ed. Leipzig, 1896. *The So-Called Historical Jesus and the Historical Biblical Christ*. Translated by Carl E. Braaten. Philadelphia: Fortress, 1964.

Kee, Howard Clark. *Community of the New Age: Studies in Mark's Gospel*. Philadelphia: Westminster, 1977.

———. "The Function of Scriptural Quotations and Allusions in Mark 11–16." In E. Earle Ellis and Erich Grässer, eds., *Jesus und Paulus: Festschrift für Werner Georg Kümmel*. Göttingen: Vandenhoeck and Ruprecht, 1975, 165–88.

Kelber, Werner H. *The Oral and the Written Gospel: The Hermeneutics of Speaking and Writing in the Synoptic Tradition, Mark, Paul, and Q*. Philadelphia: Fortress, 1983.

Kennedy, George. "Classical and Christian Source Criticism." In W. O. Walker, ed., *The Relationships among the Gospels: An Interdisciplinary Dialogue*, San Antonio, Tex.: Trinity University Press, 1978: 125–55.

———. *New Testament Interpretation Through Rhetorical Criticism*. Chapel Hill: University of North Carolina, 1984.

———, ed., *The Cambridge History of Literary Criticism*. Vol. 1, *Classical Criticism*. Cambridge: Cambridge University Press, 1989.

Kermode, Frank. *The Genesis of Secrecy: On the Interpretation of Narrative.* Cambridge: Harvard University Press, 1979.

Kirshenblatt-Gimblett, Barbara. "A Parable in Context: A Social-Interactional Analysis of Storytelling Performance." In D. Ben-Amos and Kenneth S. Goldstein, eds., *Folklore: Performance and Communication*, 105–30. The Hague and Paris: Mouton, 1975.

Kline, Meredith. "The Old Testament Origins of the Gospel Genre" *Westminster Theological Journal* 38 (1975): 1–27.

Koenig, John. "*St Mark* on the Stage: Laughing all the Way to the Cross." *Theology Today* 36 (1979): 84–88.

Koester, Helmut. *Ancient Christian Gospels: Their History and Development.* London: SCM; Philadelphia: Trinity, 1990.

———. *Introduction to the New Testament.* 2 vols. Berlin: de Gruyter, 1982.

———. "One Jesus and Four Primitive Gospels." In James M. Robinson and Helmut Koester, *Trajectories through Early Christianity*, 158–204. Philadelphia: Fortress, 1971.

———. "The Story of the Johannine Tradition" *STR* 36.1 (1992) 17–32.

Kloppenborg, John S. *The Formation of Q: Trajectories in Ancient Wisdom Collections.* Philadelphia: Fortress, 1987.

Knox, Bernard M. W. "Silent Reading in Antiquity." *GRBS* 9.4 (1968): 421–35.

Larson, Janet Karsten. "St. Alec's Gospel." *Christian Century* 96.1 (1979): 17–19.

Lefkowitz, Mary R. *The Lives of the Greek Poets.* Baltimore, Md.: Johns Hopkins University Press, 1981.

Leo, Friedrich. *Die griechisch-römanische Biographie nach ihrer literarischen Form.* Leipzig: Teubner, 1901.

Levison, John R. *Portraits of Adam in Early Judaism: From Sirach to 2 Baruch.* Sheffield: JSOT, 1988.

Lewis, Clive Staples. Preface to *Paradise Lost.* London: Oxford University Press, 1942.

———. *A Grief Observed.* New York: Seabury Press, 1961.

Lightfoot, R. H. *The Gospel Message of St. Mark.* London: Oxford University Press, 1950. 2d ed., corrected, 1952.

———. *History and Interpretation of the Gospels.* Bampton Lectures, 1934. London: Hodder and Stoughton, 1935.

Linnemann, Eta. *Jesus of the Parables: Introduction and Exposition.* New York: Harper and Row, 1966.

Lohr, Charles H. "Oral Techniques in the Gospel of Matthew." *CBQ* 23.4 (1961): 403–35.

Lord, Albert Bates. "The Gospels as Oral Traditional Literature." In William O. Walker, ed., *The Relationship among the Gospels: An Interdisciplinary Dialogue*, 33–91. San Antonio, Tex.: Trinity University Press, 1978.

———. *The Singer of Tales*. Cambridge: Harvard University Press, 1960.

MacMullen, Ramsey. *Paganism in the Roman Empire*. New York: Yale University Press, 1981.

Magarshack, David, ed. *Stanislavsky on the Art of the Stage*. London: Faber, 1967.

Malbon, Elizabeth Struthers. *Narrative Space and Mythic Meaning in Mark*. San Francisco, Calif.: Harper and Row, 1986.

———. "The Poor Widow in Mark and her Poor Rich Readers." *CBQ* 53.4 (1991): 589–604.

Marrou, Henri I. *A History of Education in Antiquity*. London: Sheed and Ward, 1956.

Martin, Ronald. *Tacitus*. Berkeley and Los Angeles: University of Calfornia Press, 1981.

McDonald, J. I. H.: *see* Chilton, Bruce, and J. I. H. McDonald.

McNeil, Brian. "Midrash in Luke?" *Heythrop Journal* 19 (1978): 399–404.

Meeks, Wayne A. *The First Urban Christians: The Social World of the Apostle Paul*. New Haven: Yale, 1983.

Meier, John P.: *see* Brown, Raymond E., and John P. Meier.

Melbourne, Bertram L. *Slow to Understand: The Disciples in Synoptic Perspective*. Lanham, Md.: University Press of America, 1988.

Moessner, David. P. "And Once Again, What sort of 'Essence'? A Response to Charles Talbert." In Mary Gerhardt and James G. Williams, eds., *Genre, Narratology, and Theology*. Semeia 43. 75–84. Decatur, Ga.: Scholars, 1988.

Momigliano, Arnaldo. *The Development of Greek Biography*. Cambridge: Harvard University Press, 1971.

———. *Studies in Historiography*. London: Weidenfeld and Nicholson, 1966.

Nagy, Gregory. *The Best of the Achaeans: Concepts of the Hero in Archaic Greek Poetry*. Baltimore, Md.: Johns Hopkins University Press, 1979.

Neill, Stephen. *The Interpretation of the New Testament 1861–1961*. Firth Lectures, 1962. London: Oxford University Press, 1966.

Neusner, Jacob. *The Idea of Purity in Ancient Judaism*. Leiden: Brill, 1973.

———. *In Search of Talmudic Biography: The Problem of the Attributed Saying*. Atlanta, Ga.: Scholars, 1984.

———. *Invitation to Midrash: The Workings of Rabbinic Biblical Interpretation*. San Francisco, Calif.: Harper and Row, 1989.

———. *Method and Meaning in Ancient Judaism: Third Series*. Atlanta, Ga.: Scholars, 1981.

———. *Rabbinic Traditions about the Pharisees before 70*. 3 vols. Leiden: Brill, 1971.

———. *What Is Midrash?* Philadelphia: Fortress, 1987.

Ong, Walter J., S.J. *Orality and Literacy: The Technologizing of the Word*. London and New York: Methuen, 1982.

———. *The Presence of the Word: Some Prolegomena for Cultural and Religious History*. Minneapolis: University of Minnesota, 1981 (copyright 1967, Yale University).

Parry, Adam. ed. *The Making of Homeric Verse: The Collected Papers of Milman Parry*. New York: Oxford University Press, 1971.

Parsons, Mikael C. "Reading a Beginning / Beginning a Reading: Tracing Literary Theory on Narrative Openings." In Dennis E. Smith, ed., *How Gospels Begin*, 11–32. Semeia 52. Atlanta, GA: Scholars, 1991.

Parunak, H. Van Dyke. "Oral Typesetting: Some Uses of Biblical Structure." *Biblica* 62 (1981): 153–68.

———. "Transitional Techniques in the Bible." *JBL* 102.4 (1983): 525–48.

Pelling, Christopher B. R."Childhood and Personality in Greek Biography." In Christopher Pelling, ed., *Characterization and Individuality in Greek Literature*, 213–44. Oxford: Clarendon, 1990.

———. "Plutarch's Adaptation of His Source Material." *JHS* 100 (1980): 127–40.

———. "Plutarch's Method of Work in His Roman Lives." *JHS* 99 (1979): 74-96.

Perdue, Leo G. "The Death of the Sage and Moral Exhortation: From Ancient Near Eastern Instructions to Graeco-Roman Paraenesis." In Leo G. Perdue and John G. Gammie, eds., *Paraenesis: Act and Form*, 81–109. Semeia 50. Atlanta, Ga.: Scholars, 1990.

Perry, Ben Edwin. *The Ancient Romances: A Literary-Historical Account of Their Origins*. Berkeley and Los Angeles: University of California Press, 1967.

Perry, Menakhem. "Literary Dynamics: How the Order of a Text Creates Its Meaning." *Poetics Today* 1 (1979): 35–64, 311–61.

Peterson, Norman R. *Literary Criticism for New Testament Critics*. Philadelphia: Fortress, 1978.

Propp, Vladimir. *Morphology of the Folktale*. 2d ed., rev. Austin: University of Texas Press, 1968.

Reardon, B. P. "Aspects of the Greek Novel." *Greece and Rome*, 2d ser., 23.2 (1976): 118–31.

———. *The Form of Greek Romance*. Princeton: Princeton University Press, 1991.

Rhoads, David, and Donald Michie. *Mark as Story: An Introduction to the Narrative of a Gospel*. Philadelphia: Fortress, 1982.

Ricouer, Paul. "Interpretative Narrative." In Regina Schwartz, ed., *The Book and the Text: The Bible and Literary Theory*, 237–57. Oxford: Basil Blackwell, 1990, pp. 237–57.

Rimmon-Kenan, Shlomith. *Narrative Fiction: Contemporary Poetics*. New York: Methuen, 1983.

Robbins, Vernon K. "The Chreia." In David A. Aune, ed., *Greco-Roman Literature and the New Testament*, 1–23. Atlanta, Ga.: Scholars, 1988.

———. *Jesus the Teacher: A Socio-Rhetorical Interpretation of Mark.* Philadelphia: Fortress, 1984.

———, and Burton L. Mack. *Rhetoric in the Gospels: Argumentation in Narrative Elaboration.* Philadelphia: Fortress, 1987.

Roberts, Colin H., and T. C. Skeat. *The Birth of the Codex.* London: Oxford University Press for The British Academy, 1987.

Sanders, E. P. *Paul and Palestinian Judaism: A Comparison of Patterns of Religion.* London: SCM, 1977.

———, and Margaret Davies. *Studying the Synoptic Gospels.* London: SCM, 1989.

Sayers, Dorothy. *The Mind of the Maker.* London: Methuen, 1941.

Schmidt, Karl L. *Der Rahmen der Geschichte Jesu: Literarkritische Untersuchungen zur Ältesten Jesusüberlieferung.* 1919 (reprint, Darmstadt: Wissenschaftliche Buchgesellschaft, 1964).

Schwartz, Regina, ed. *The Book and the Text: The Bible and Literary Theory.* Oxford: Basil Blackwell, 1990.

Selvidge, Marla J. *Woman, Cult, and Miracle Recital: A Redactional Critical Investigation on Mark 5:24–34.* Lewisburg: Bucknell University Press; London and Toronto: Associated University Press, 1990.

Senior, Donald, C. P. *The Passion of Jesus in the Gospel of Mark.* Wilmington, Del.: Glazier, 1984.

Shuler, Philip L. *A Genre for the Gospels: the Biographical Character of Matthew.* Philadelphia: Fortress, 1982.

Silberman, Lou H. "Reflections on Orality, Aurality and Perhaps More." In L. H. Silberman, ed., *Orality, Aurality and Biblical Narrative.* Semeia 39. Atlanta, Ga.: Scholars, 1987, pp. 1–6.

Standaert, B. H. M. G. M. *L'Evangile selon Marc: Composition et Genre Littéraire.* Zevenkerken: Brugge, 1978.

Stock, Augustine. O.S.B. *Call to Discipleship: A Literary Study of Mark's Gospel.* Wilmington, Del.: Glazier, 1982.

———. "Hinge Transitions in Mark's Gospel." *Biblical Theology Bulletin* 15 (1985): 27–31

———. *The Method and Message of Mark.* Wilmington, Del.: Glazier, 1989.

Talbert, Charles H. "Once Again: Gospel Genre." In Mary Gerhardt and James G. Williams, eds., *Genre, Narratology, and Theology*, 53–73. Semeia 43 (Scholars, 1988).

———. *What Is a Gospel? The Genre of the Canonical Gospels.* Philadelphia: Fortress, 1977.

Taylor, Vincent. *The Formation of the Gospel Tradition.* London: Macmillan; New York: St. Martin's, 1933.

———. *The Gospel according to St. Mark.* London: Macmillan, 1952.

Tolbert, Mary Ann. *Sowing the Gospel: Mark's World in Literary-Historical Perspective.* Minneapolis, Minn.: Fortress, 1989.

Turner, C. H. "Marcan Usage: Notes, Critical and Exegetical, on the Second Gospel." *JTS* 26 (1925): 145–46.

Ulansey, David. "The Heavenly Veil Torn: Mark's Cosmic Inclusio." *JBL* 110.1 (1991): 123–25.

Urbach, Ephraim. *The Sages: Their Concepts and Beliefs*. Jerusalem: Magnes, 1975. 2nd ed., enlarged, 1979.

Vermes, Geza. *Scripture and Tradition in Judaism*. Leiden: Brill, 1973.

Via, Dan O. *The Ethics of Mark's Gospel: In the Middle of Time*. Philadelphia: Fortress, 1985.

———. *The Parables: Their Literary and Existential Dimension*. Philadelphia: Fortress, 1967.

Vinaver, Eugène: *see* Primary Sources, Sir Thomas Malory.

Voelz, J. W. "The Language of the New Testament." *ANRW* 25.2 (1984): 893–977.

Votaw, C. W. "The Gospels and Contemporary Biographies." *American Journal of Theology* 19 (1915): 45–73, 217–49.

Waetjen, Herman C.. *A Reordering of Power: A Socio-Political Reading of Mark's Gospel*. Minneapolis, Minn.: Fortress, 1989.

Wallace-Hadrill, Andrew. *Suetonius: The Scholar and His Caesars*. London: Gerald Duckworth, 1983; New Haven: Yale University Press, 1984.

Wellek, René, and Austin Warren. *Theory of Literature*. 3d ed. Harmondsworth, Middlesex: Penguin, 1963.

Woude, Simon Adam van der. "Messianic Ideas in Later Judaism." Part 4 in Gerhard Friedrich, *Theological Dictionary of the New Testament,* vol. 9, 517–20. Grand Rapids, Mich.: Eerdmans, 1974.

Wright, Addison G. "The Widow's Mite: Praise or Lament?—A Matter of Context." *CBQ* 44.2 (1982): 256–64.

Zumthor, Paul. *Oral Poetry: An Introduction*. Minneapolis: University of Minnesota, 1990.

Zwettler, Michael. *The Oral Tradition of Classical Arabic Poetry: Its Character and Implications*. Columbus: Ohio State University Press, 1978.

AUDIO-VISUAL RESOURCE

McCowan, Alec. *St. Mark's Gospel: King James Version*. Videotape. Distributed by the American Bible Society, 1865 Broadway, New York, NY 10023.

Rhoads, David. *Dramatic Performance of Mark*. Videotape. Distributed by Select, 2199 E. Main Street, Columbus, Ohio.

Index to Modern Authors

Index to Ancient Sources

Subject Index

Scriptural Index

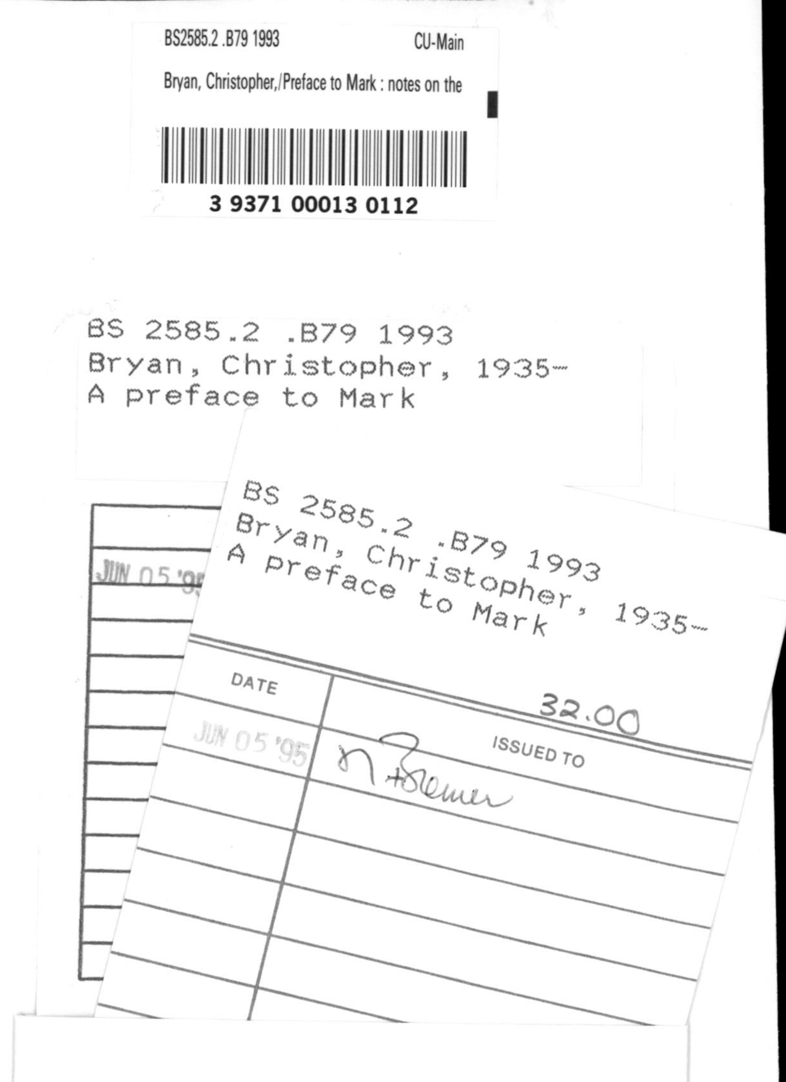
BS 2585.2 .B79 1993
Bryan, Christopher, 1935-
A preface to Mark
JUN 05 '95
BS 2585.2 .B79 1993
Bryan, Christopher, 1935-
A preface to Mark
DATE
ISSUED TO
32.00
JUN 05 '95